W. FRED CONWAY

MORE FIREFIGHTING LORE

**Forty more strange but true stories
from firefighting history, including
*"The Mississippi Steamboat Fire
Whose Death Toll Exceeded The Titanic"***

Foreword by HAL BRUNO

**Political Analyst with ABC News in Washington,
and a contributing editor of *Firehouse Magazine*®**

Library of Congress Cataloging in Publication Data
Conway, W. Fred
More Firefighting Lore
Library of Congress Catalog Number: 58-093030
ISBN 0-925165-23-9

Fire Buff House Publishers
P.O. Box 711, New Albany, IN 47151-0711
© W. Fred Conway 1998.
All rights reserved.

Printed in the United States of America
Typography and Layout: Pam Campbell-Jones

10   9   8   7   6   5   4   3   2   1

Acknowledgments

Many firefighting historians have contributed material for the stories that appear on these pages. Some of them contacted me with ideas for additional stories after having read my original *Firefighting Lore*. I am grateful for their valuable contributions.

Additional stories have been gleaned from books that have been long out of print. Still other stories have come from magazines whose pages were yellowed and crumbling. All the authors and suppliers of facts for these unusual firefighting stories have my deep appreciation, and I have endeavored to list them all under "Sources" beginning on page 171. If anyone has been omitted, it is totally unintentional.

Special thanks to Dr. Tom Scott of Oak Ridge, Tennessee, whose extensive library of over 2000 fire-oriented books was once more made available for my research. Both Tom and I believe that firefighting history should be perpetuated and shared. Hopefully this book will contribute to the accomplishment of that goal.

The expertise, cooperation, and enthusiasm of the three ladies whose efforts helped bring this book to fruition are greatly appreciated: Editor – my wife Betty; Typesetting and Layout – Pam Campbell-Jones; Proofreader – Evelyn Raymer.

This book was fun to write, and I hope you'll find it fun to read these strange but true stories from the annals of firefighting history.

W. Fred Conway
New Albany, Indiana

Foreword

The American fire-rescue service is rich in tradition and legends. Its history and lore go back to the days of the first bucket brigades and the volunteers of the early 19th century with their ornately decorated, hand-pumped "masheens." It continues through the 1800's, with the first paid departments and those magnificent horse-drawn steamers; and into the 20th century, with the first frail-looking motorized apparatus that has evolved into the monstrous engines, ladder trucks and rescue squads we know today.

A lot has changed since 1735, when Benjamin Franklin organized the Union Hose Company in Philadelphia. But much has remained the same — the fascination we have for the apparatus, some of the tools that are still being used and, most importantly, the courage and devotion to duty of America's firefighters. As we enter the 21st century, the spirit of our fire companies is as strong as it ever has been, whether it's volunteers in a small town or career firefighters in a big city; whether they're responding to a fire, an accident or an emergency medical call. Whatever the disaster may be, our firefighters will respond, as they have for more than 260 years.

In this book, W. Fred Conway once again has captured that spirit with more legends and lore of the American fire service. These are the stories firefighters and fire buffs talk about with each other as they share a common bond that is the historic link from the present to the past. There are stories of comedy and tragedy; together, they are the fabric of every firefighter's life.

It has been my life since childhood. I first heard fire stories from my father, who grew up as a poor boy on New York's Bowery in the early 1900's. He became a buff at Engine 33 on Great Jones Street and, at one point, the firemen cared for him in the firehouse and saved him from being sent to an orphanage. He told me about the Boston Excelsior, Triangle Shirtwaist, Equitable Building and other terrible fires that Engine 33 fought in the area known as "Hell's Hundred Acres." And, over the years, I heard more stories of firefighters in other places caring for children who needed help. That is a little known part of firehouse lore and it deserves to be told.

I was attracted to the fire department as a boy in Chicago, hanging out in the firehouses, running errands, chasing after the apparatus when they got a run, and traveling to every part of the city to help roll up the hose at the big extra-alarm fires — often earning a ride back to a firehouse aboard an engine. Eventually, I became a volunteer firefighter and, over a period of 40 years, served with five different fire departments as my work in journalism caused me to move around the country.

But I've always remembered the tales I heard from those leather-lunged firemen in the Chicago Fire Department. Therefore, it's only proper that I contribute one of those CFD stories to Fred Conway's rich treasure of firefighting lore. It goes like this...

Here Comes der Captain

Hook-and-Ladder 10, on Chicago's near north side, was once commanded by a captain who had immigrated from Germany and ran his firehouse like a Prussian drill school. He was a good fire officer, but a strict disciplinarian, who spoke with a thick Germanic accent and had a habit of referring to himself in the third person. One day he ordered Truck 10 into the alley next to the firehouse to practice with their life net and, after reviewing the correct way to use a net, announced: "Der Captain vill go up to der vindow and jump into der net you are holding."

But the phone rang as he walked into the firehouse and the captain got involved in a long conversation. Suddenly, he remembered that his crew was waiting in the alley. He slammed down the phone, dashed up the stairs, ran across the bunk room and leaped from the open window. Unfortunately, the firemen had grown tired of waiting and had put the life net down. They were leaning against a fence when they heard a noise and looked up in horror to see their leader sailing out of the window in a perfect sitting position, as he shouted: "Here comes der Captain!"

He landed in the alley with a bone-breaking thud; it was many months before he recovered from his injuries and returned to duty. But he became a legend and for years afterward, whenever anything went wrong, a firefighter would simply say, "Here comes der Captain" — and everyone in the Chicago Fire Department knew exactly what that meant.

— Hal Bruno
Political Analyst with ABC News in Washington, and a contributing editor of Firehouse Magazine®

Table of Contents

The Mississippi Steamboat Fire Whose Death Toll Exceeded The *Titanic*

It was a disaster waiting to happen, and it did. Overloaded with more than 2,100 Union soldiers who had just been released from the infamous Andersonville and Cahaba Confederate prison camps during April, 1865, the two-year-old 260-foot-long steamboat *Sultana* took 1,700 of them to firey deaths.

The prisoners, many of them emaciated and sick from long months of depravation in prison, boarded the *Sultana* four miles west of Vicksburg, Mississippi, thankful that they had survived the war and would soon be with their families and loved ones. But it was not to be.

The *Sultana* was scheduled to travel up the Mississippi River to Cairo, Illinois, and then up the Ohio River toward Camp Chase in Ohio, where the men would be mustered out.

The co-owner of the *Sultana,* Captain J. Cass Mason, was in such financial difficulty that he had sold a part interest in the boat, but he was still deeply in debt. Learning that he could receive $5.00 for each enlisted man and $10.00 per officer for each paroled prisoner of war he could haul north, he conspired to haul just as many as he possibly could.

The April 12, 1865, Inspector's Certificate certified the *Sultana* as being safe. Yet, just two weeks later, defective boilers resulted in the explosion and fire that took between 1,700 and 1,800 lives – exceeding the death toll of the *Titanic*.

The legal carrying capacity of the *Sultana* was 376, so with her crew of 86, she could legally carry only 290 prisoners. But, ignoring legalities, and blinded by the financial windfall he would receive, Captain Mason crammed ailing prisoners onto the *Sultana* like sardines, and boarded over 2,100 of them. When completely loaded, the *Sultana* held 86 crewmen, 100 civilian passengers, and more than 2,100 soldiers. In addition, 200 head of horses, mules, and hogs, as well as the Captain's pet alligator were on board. Captain Mason even went so far as to unload 250,000 pounds of sugar to make room for higher paying human cargo.

With the ballast removed and the upper deck crammed with just-released prisoners of war, the *Sultana* became top heavy, which was one of the contributing factors to the explosion and fire that took more than 1,700 lives, but there were other factors as well.

On April 12, just two weeks before the disaster, government inspectors in St. Louis reported that the *Sultana* "may be employed as a steamer upon the waters herein specified without peril to life from imperfection of form, materials, workmanship, or arrangement of the several parts or from age or use." But the inspectors notwithstanding, crew members were concerned about the steamer's massive boilers.

One crew member knew that the boilers, little more than two years old, had already been patched at both Vicksburg and Natchez, Mississippi, on two previous trips. As the *Sultana* drew within several miles of the parole camp near Vicksburg, where she would onload the ill-fated prisoners, steam was discovered escaping from a crack in one of her boilers.

A Vicksburg boilermaker, R.G. Taylor, was called in, and advised that extensive repairs were needed. But Captain Mason knew that if the *Sultana* did not leave Vicksburg on April 24th, loaded with prisoners, that some other steamboat would reap the financial windfall that he so desperately needed. He begged Taylor to do a temporary patch job. Reluctantly, Taylor finally agreed after Mason promised to have permanent repairs made as soon as the boat reached St. Louis. But the *Sultana* never made it that far.

The boiler patch was too thin, and with the boilers operating at full pressure, the patch simply couldn't hold.

The *Sultana's* four boilers were configured with tandem piping so that if the boat tilted, the water from one boiler would run to the next. Being top heavy, with her 376 person capacity now up to more than 2,500 – more than six and one-half times her legal capacity – she tilted at about 2:00 a.m. on April 27th, near Memphis, Tennessee. The water from the end boiler drained into the other boilers, exposing the first boiler's metal to direct heat since no water was against it.

When the boat righted itself, the water drained right back into the first boiler, which caused an immediate increase in steam and a buildup of pressure. The temporary patch blew out with the force of a gunpowder explosion, tearing apart the furnaces beneath the boilers and scattering hot coals. Soon the *Sultana* was a mass of flames.

As the flames leaped skyward, the men who had survived the explosion leaped into the swollen four-mile-wide Mississippi River. Hundreds of them, after

Hundreds of emaciated Civil War prisoners just released from Andersonville and Cahaba Confederate prisons struggle in the water after the steamboat bringing them home exploded and caught fire.

Woodcut from *Harper's Weekly*

surviving the rigors of Confederate prison camps, drowned in the cold waters of the raging Mississippi.

The Captain's Pet Alligator To The Rescue

The following account is by one of the *Sultana* survivors, Jesse Huffaker:

"The river was out of its banks. Mush ice was running in the river, which numbed our bodies. Some swam ashore. One man rode out on a dead horse. There was an alligator on the boat. A man in the water was about to go under. He heard a noise behind him. Thinking it was the alligator, with renewed energy he swam ashore."

Yet, more than 700 did make it ashore and were taken to Memphis hospitals and the Memphis Soldiers' Home. But many were so severely scalded by steam, badly burned, or injured that their misery was of short duration, being relieved by death within hours.

Although Captain Mason survived the blast and was seen helping the men, he did not live to face charges and was presumed burned or drowned. The fire finally died down, somewhat of its own accord, and the 300 men still on board were rescued by men rowing out from the shore. As the last soldier was removed from the stricken boat, the *Sultana* sank beneath the muddy waters of the Mississippi.

The death toll of 1,547 did not include some 200 injured and burned who died in Memphis hospitals during the next few days. The total count was over 1,700 and possibly as many as 1,800.

Nearly a half century later the *Titanic* sank, claiming 1,517 lives. But the now almost forgotten flaming *Sultana* exceeded the *Titanic* disaster in number of lives lost by more than 200.

The Fire Inventor Whose Tanks Helped Win World War II

J. Walter Christie, whose invention of the motorized tractors that replaced thousands of fire horses in the early 1900's, is well known to many fire buffs. But perhaps less well known is the fact that those tractors financed his foray into the design of many of the Allied tanks used in World War II.

Born in 1886 on a farm in New Jersey, Christie was a natural-born mechanical genius. At age 12 he whittled a boat from a wooden log and propelled it with parts from his family's grandfather clock, which he had removed surreptitiously. To escape his father's wrath, he hid at his cousin's house a mile away for several days. At age 16 he left home to seek his fortune. Before his life ended, he had made and lost several fortunes.

His patent for 'Piston Packing For Steam Engines,' which he invented when he was 20 years old, established him as a first class inventor, and other accomplishments soon followed – an underground heating system for New York City buildings, overhead money chutes for New York Department stores including Gimbels, Wannamakers and Macys, and a mechanical gearing assembly to raise and lower operating tables at the Mayo Clinic.

But as the Twentieth Century dawned, Christie's inventive talents took a new turn. He became consumed with not only designing and building, but driving race cars. He consistently won races, attaining speeds of 100.5 mph in 1905, and 120 mph in 1906. But after being seriously injured in a racing accident, he decided to build taxi cabs.

The taxi business failed when his financial backer disappeared, which together with his racing expenses drained his finances to the limit. He sold his fastest race car in which he had invested over $50,000 to legendary race car driver Barney Oldfield for a mere $750. He was down on his luck.

Down, but not out, Christie conceived the idea for the motorized tractor to pull steam fire engines. Soon he found a wealthy financier to underwrite the cost of a new factory in which to build them. Said the new-found financier, "Mr. Christie, I am aware of the possibility of your Front-Wheel Drive being a great success. I have watched your fabulous racing cars win many races. As you progress with your factory, I shall advance additional funds."

The factory was completed, and the first front wheel drive tractor to pull fire apparatus was demonstrated before mayors, fire commissioners, and city officials from across the country. The demonstration was a howling success, and the media had a field day. Orders for the tractors poured in. Boston Mayor James Curley was so impressed with the tractors he had purchased that he said, "The Boston Fire Department's only complaint was that they had not bought more of them."

Soon the factory had to be expanded to handle all the orders, and by 1918 over 600 of the tractors were in service in fire departments throughout America. Christie's up and down fortunes had soared again.

Then, incredibly, at the peak of his new success, Christie began to turn down orders for more tractors. He had begun to invest his new found fortune, his time, and his talent in still another invention – the first American Army tank.

He had received a War Contract, and the majority of his workers were shifted from producing fire engine tractors to building military vehicles. His bonding company tried to force him to pay forfeits for non-delivery of fire engine tractors, but Christie convinced them that war conditions (World War I) took precedence over prior commitments, and he was absolved of any responsibility.

In 1918, the United States Army did not have its own tank design, having borrowed the designs of the British and French. The "Christie Tank" outperformed them both. But the war was soon over, and tank orders ceased. Christie's profits from the fire engine tractors had been invested in tank production, and once again he was out of funds.

Somehow he managed to stay solvent, and his factory survived the depression years. By selling some of his patents, he raised the money for his next invention – a High Speed Tank – which was revolutionary in design. When the U.S. Army at first spurned his new design, he was approached by various foreign governments, including Japan, but to his credit, he held out, hoping that his own country would finally buy it.

One of some 33,000 Russian-built Christie tanks built between 1931 and 1938.

And it did! In 1929, after two frustrating years of rejections, the Chief of Ordnance to the Assistant Secretary of War stated, "I am quite in favor of applying the $250,000 appropriated for light tanks for the purchase of Christie vehicles."

U.S. Army Major C.C. Benson gave the government a report which said, "Mr. Christie has given us in his new machine more technical progress than Great Britain has secured during the past ten years from the expenditure of some sixty-five million dollars."

Col. Cooper, Commandant of the Tank School, commented, "I will say unhesitatingly that it is so far superior to anything else we have in the tank line, that there is no comparison."

Major Brett, Executive Officer of the Mechanized Force, reported, "It is unquestionably the most attractive thing that there is in any tank service in the world at the present time."

Christie and American LaFrance Meet Again

For at least two decades – from 1910 to 1930 – Christie tractors pulled hundreds of steam fire engines which had been manufactured by the various companies who consolidated in 1902 to form American LaFrance, America's premier fire engine manufacturer.

Fate decreed in 1932 that American LaFrance and Christie would be teamed in an entirely different way.

After purchasing the first seven Christie tanks, the U.S. Army decided that they wanted five more, but they allowed that they now owned Christie's design. The five additional tanks, to be built to Christie specifications but with modifications, were not ordered from Christie, but were put out for bid.

Christie refused to bid because he knew that the Army's modifications would overload the tank, reducing its performance. The winning bidder was none other than American LaFrance!

But, just as Christie had predicted, the tanks failed to pass the test and were rejected. This caused a considerable financial loss to American LaFrance, who bounced back to producing fire apparatus, which they continue to do to this day.

This Christie tank was built in 1932 by American LaFrance.

General Fuqua, Chief of Infantry, stated succinctly, "It is the best [tank] in the world today."

None other than General Douglas MacArthur agreed. He said, "This is the best tank that has ever been developed."

Yet, despite all these accolades, the United States Army ordered only seven of the tanks. Desperate for funds, Christie sold two of them to Russia. As a result, during World War II, some 33,000 Russian-built Christie tanks helped to beat back the German army at Stalingrad, the battle that turned the tide of the war against Germany.

Earlier, learning of the Christie tank, Adolph Hitler had sent his personal representative, a banker named Stinnes, to offer Christie one million dollars to supervise construction of his tanks in Germany. To his eternal credit, Christie refused in the presence of his son, who said, "Dad, it took guts to turn down such an easy fortune." "I know, son," Christie replied, "but it can't buy your soul, and that's about all I have left. One day we'll be at war with Hitler."

Now, down and out for the last time, Walter Christie was too old to build another fortune.

But, today, Great Britain owes a debt of gratitude to Christie, for it was the tanks he designed that enabled the British to defeat German General Rommel in North Africa, a fact acknowledged by Winston Churchill. Additionally, Christie-designed tanks had prevented the Germans from capturing Moscow.

On January 11, 1944, Walter Christie passed away at age 68, without even the money to pay his doctor bills.

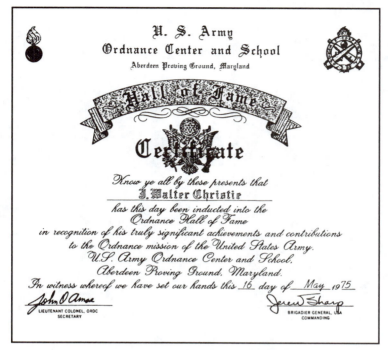

Posthumously, in 1975, J. Walter Christie was inducted into the Ordnance Hall Of Fame of the United States Army, Aberdeen Proving Ground, Maryland. His country, which had scorned him in life, honored him three decades after his death "...in recognition of his truly significant achievements and contributions to the Ordnance mission of the United States Army."

America's Foremost Female Firefighter

The Coit Tower, today a San Francisco landmark, was built atop Telegraph Hill to honor the memory of the illustrious lady considered to be America's first female fire buff, the winsome but boisterous Lillie Hitchcock Coit.

Born in 1843, Eliza Wychie Hitchcock was never known by her given name. Soon after her birth, her Army doctor father declared that she was as fair as a lily, and "Lillie" she became. "She is the merriest, sweetest creature in the world, always laughing and frolicking," read an entry in her mother's diary.

When Lillie was seven, her father was appointed Medical Director of the Pacific Coast, and the family moved from North Carolina to San Francisco, a city bustling with Gold Rush activity. Lillie, her father, and her mother took up residence at the Howard Boarding House on Telegraph Hill. Lillie was visiting with friends one day when she heard the fire bell ring and saw flames racing up Telegraph Hill, consuming everything in their path, including the Howard Boarding House.

Their temporary home destroyed, the family moved into the Oriental Hotel, and Lillie, along with newfound

Fire Belle Lillie Hitchcock Coit, the little girl who grew up to become a fire service legend.

friends Patrick and Joana Fitzmaurice, was playing on the second floor of another nearby hotel which was still under construction, when fire suddenly raced up the stairs blocking their escape.

Lillie headed for the rear stairs, but her friends refused to follow her and were burned to death. With the rear stairs burning as well, Lillie again found herself trapped. She was petrified. What happened next was the incredible rescue that Lillie would never forget. Volunteer fireman John Boynton of Knickerbocker No. 5 Engine Co. chopped a hole in the roof, let himself down on a rope, ordered Lillie to jump on his back, then pulled himself, with Lillie clinging to him, back up the rope to the roof, where they descended a fire ladder to the ground. That baptism by fire forever changed Lillie's life.

Her rescuer, John Boynton, responded to another fire the following night, and he recalled, "As we dashed past the Oriental, I saw the bright-eyed piquant little girl I'd rescued the day before. She was standing on the piazza of the hotel holding fast to her mother's hand. As we swept closely by on the narrow street, she called out to us, 'Hurrah for my dear Number Five.' "

"That night," he continued, "when we returned to the fire hall, we found a barrel of brandy sent to us by her father, Dr. Hitchcock, and a thousand dollars toward a new engine. But it was Lillie's heart which throbbed with eternal love for the members of No. 5. From then on, she belonged to us as much as we belonged to her."

When the fire bell rang and the volunteers of Knickerbocker No. 5 pulled their piano box hand engine to the fire by its drag ropes, it was common for young boys in the neighborhood to fall in behind the engine to join the excitement. Never had a girl joined in. Although Lillie was forbidden by her parents to do so, some things are so irresistible that parental authority is flouted.

The wild ways of Lillie were becoming a legend among all 14 volunteer fire companies in San Francisco.

Lillie was eight years old when the fire bell rang for another fire on Telegraph Hill as she was on her way home from school. Throwing her books to the ground, she not only joined the procession, but as the volunteers struggled to drag the engine up the steep hill, she took her place on the rope and pulled with all her might. Bystanders watching the little girl help the firemen felt compelled to grab the rope as well, and No. 5 got up the hill so fast that they had the fire out before rival fire companies arrived.

But Lillie was just getting warmed up. Her next adventure occurred when, outfitted in a blue silk party dress, she was being driven to a playmate's birthday party in a carriage her father had hired. Seeing Knickerbocker No. 5 speeding by on the way to a fire, she ordered the driver to follow the engine. She hopped out at the fire, seized the hose, and with cinders falling on her party dress, directed the water at the flames.

The wild ways of Lillie were becoming a legend among all 14 volunteer fire companies in San Francisco. The happy little fire belle had won their hearts. She would respond to every fire she possibly could. If she couldn't help fight the flames, she would stand back and yell, "Hurrah for Number Five!" and "Pull steady, boys!" If Lillie failed to show up at a fire, the volunteers missed her and would ask, "Where's our girl? Why isn't our mascot here?"

The years passed, and the little tomboy grew into a lovely, graceful, amiable teenager, still as fair as a lily. But her haughty mother was chagrined at Lillie's fire chasing, and sent her away to a private school at San Jose. "You are making a spectacle of yourself. We are southern aristocrats, and you are shaming us," her mother declared. Soon Lillie's outrageous conduct at the school got her expelled, and back she came, still sneaking away when the fire bell rang.

When she fell ill, her friends at Knickerbocker No. 5 sent her a small gold fire helmet as a gift. As she began to recover, they gathered outside her window to serenade her. Thrilled by their attention and concern, she soon got well.

Even rival fire companies honored her by allowing her to be seated next to the driver of Knickerbocker No. 5's engine in all the frequent firemen's parades.

Even rival fire companies honored her by allowing her to be seated next to the driver of Knickerbocker No. 5's engine in all the frequent firemen's parades. The first time Howard Coit, a tall well-built young man with brown hair and twinkling green eyes saw her, she was seated on the engine. In time, Howard, 22, attended a ball where the belle of the ball was Lillie. They danced, talked, and kissed. She wrote in her diary, "This is he."

But, according to Lillie's mother, Howard was not good enough for her, even though his father was a doctor, and Howard himself had amassed a small fortune as part owner of a mine. To keep Lillie from both Howard and her fire chasing, her mother took Lillie to France. Before she left, Lillie visited the fire hall. "Goodbye dear Number Five," were her parting words. She left both of her loves – Number Five and Howard. "Howard Coit is my secret love," she confided in her diary.

In 1863, Lillie wrote to her father, pleading to come home. Her father arranged it and wrote back to say that the firemen, the mayor, the city council, and even the merchants of San Francisco would turn out to welcome her with flags flying. A newspaper stated, "The city has not been the same without Lillie."

Now our story takes still another turn. When her mother learned that she was again surreptitiously dating Howard, as well as responding to fires with Knickerbocker No. 5, she whisked her back to France. Meanwhile her father, an old friend of Jefferson Davis, was appointed as Medical Director of the Confederate Army. Lillie and her mother stayed in France until the war's end, and upon their return Lillie received a shock.

Her beloved Knickerbocker No. 5 fire company had been disbanded, having been replaced by paid firemen with a steam fire engine. Engine No. 5 had been sold to the fire department at Virginia City and was ready to be delivered, pulled by a team of six horses. Guess who drove the horses – yes, it was Lillie!

The Virginia City firemen held a banquet for the legendary Lillie and asked her to speak. She recounted many tales of Number Five and ended by saying, "I have one regret in life: that I cannot ride to another fire." At that moment an alarm sounded (it had been prearranged by the Virginia City Fire Department), and Lillie drove Number Five for the last time.

Arriving back in San Francisco, Lillie eloped with Howard, much to her mother's dismay. In time, Howard became extremely successful as Chairman of the Board of the San Francisco Stock Exchange, and Lillie's fame as a socialite did not diminish. When the new City Hall was dedicated, Lillie's photograph was placed in the cornerstone as "the most outstanding woman in San Francisco and the only lady fireman."

Howard Coit died unexpectedly of a bad heart at the early age of 47, leaving Lillie with a sizeable estate. She never remarried, and at her death in 1929 at age 88, she was worth over a million dollars. She left a bequest to the City of San Francisco of $100,000 "to beautify the city."

On October 8, 1933, old Engine No. 5, retrieved from Virginia City for the occasion, was pulled by horses up Telegraph Hill to dedicate the Coit Monument and Observation Tower erected by the city to perpetuate the memory of "Firebelle Lillie."

The Day An Airplane Hit The Empire State Building

Its 102 stories towering 1,258 feet above Manhattan, the Empire State Building on that foggy morning in 1945 was the tallest building in the world.

Built in 1930-31, it accommodated 80,000 persons, although it contained only 35,000 persons on an average day. Only around 2000 persons were in the building at 9:55 a.m. on Saturday, when a converted B-25 Mitchell bomber, flying at 250 miles per hour, slammed into the 79th floor.

The 102-story Empire State Building was the tallest structure in the world in 1945, the year it was struck and set afire by an Army Air Force bomber.

With its name *Old Feather Merchant* emblazoned under its cockpit, the bomber had just been retired from World War II service and had been converted to carry passengers. Taking off at 8:55 a.m., exactly one hour before the crash, from the Bedford Army

As the Army Air Force bomber slammed into the 79th floor, high octane aviation gasoline spread through elevator shafts, starting a four-alarm fire.

Airfield near Boston, was experienced pilot Lt. Col. Bill Smith, with SSgt. Christopher Domitrovich, the crew chief, acting as co-pilot. A young sailor, Albert Perna, was a passenger, having hitched a ride to New York City on the ill-fated plane. Never could they have imagined that within the next hour their lives would be snuffed out by a collision with – of all things – the Empire State Building.

Working with the New York City Fire Department, architect William Lamb had designed the most modern and sophisticated fire protection system ever installed in an office building up to that time. Throughout the mighty structure were 400 fire hose connections on an eight-inch standpipe system fed by six tanks on various building levels, each holding 10,000 gallons of water. Otherwise, to pump water to the top floor would have required the impractical pressure of 542 pounds per square inch. In addition, the building was totally sprinklered and central station monitored.

At 9:43 a.m. the flight ceiling had dropped sharply, and pilot Col. Smith now had zero visibility. As he approached LaGuardia Field, he put the B-25 into a shallow dive down to 650 feet, which was below the ceiling, and he could again see. But seven commercial airliners were stacked up in a holding pattern within 25 miles of LaGuardia, all flying blind, waiting for clearance to land. *The Old Feather Merchant* was diverted to Newark. Somehow Smith became confused when he spotted Welfare Island, mistaking it for Manhattan. Now he lost altitude to just 550 feet and lowered his landing gear. It was 9:54 a.m.

Patrolman Albert Schneider was walking his beat on Lexington Avenue when he heard a thunderous roar overhead. "He's heading straight for Radio City!" Schneider yelled to a nearby delivery man, "and he ain't gonna make it!"

> *In the split second before the impact, Smith, Domitrovich, and Perna could see people seated at their desks on the 79th floor.*

In a desperate attempt to regain altitude, Col. Smith got the plane up to 670 feet, just clearing a 600-foot tall building. A 700-foot building was at the corner of Fifth Avenue and 42nd Street. Now up to 710 feet, Smith cleared it by a scant ten feet. Still climbing, he continued past the 725-foot tall Chrysler Building with no room to spare.

He was now seven blocks from the 1,258-foot-tall Empire State Building, traveling at 250 miles per hour. He pulled the column control into his chest, putting the bomber's nose into an almost vertical position, with the right wing perpendicular to the face of the Empire State Building. In the split second before the impact, Smith, Domitrovich, and Perna could see people seated at their desks on the 79th floor.

As the sound of the explosive thud echoed across the Hudson River, the clouds enshrouding the Empire State Building suddenly parted, exposing an arrow of orange flames streaking out from the north face of the structure. With World War II still not over, some thought it was a Japanese bomb, while others feared a German V-2 rocket. A patrolman opened a street call box and exclaimed, "This is Kennedy – Badge 3772. I think a plane just hit the Empire State Building!"

The initial response brought sixteen engines, five hook and ladder trucks, and a rescue squad. Soon twenty-three companies were at the scene in command of legendary Fire Chief Patrick Walsh. Firemen carried rolls of 2-1/2 inch hose into operating elevators, which stopped at the 60th floor. Then they carried the heavy double-jacketed cotton-lined rubber hose up another 18 flights of stairs to the floor beneath the crash site and connected the hose to the standpipe system. The standpipes were intact. There was pressure!

Meanwhile, other firemen had gone to the sub-basement, where two elevator cars had embedded themselves three to four feet deep into the concrete floor after having fallen over 1000 feet. No one could have possibly survived, but wait ... was that a moan coming from one of the cars?

Elevator operator Betty Lou Oliver miraculously survived the 1,000 foot plunge when one of the plane's engines severed the cables to her elevator, sending it hurtling from the 79th floor to the sub-basement.

Elevator operator Betty Lou Oliver had miraculously survived the plunge from the 79th floor and was semi-conscious, although her back and both legs were broken. Incredibly, she completely recovered within eight months. But eleven others on the 79th floor, as well as the three occupants of *The Old Feather Merchant* perished.

Flames erupt from the building moments after an airplane slammed into the 79th floor.

Doing a heroic job, the New York City Fire Department had the fire totally out within 40 minutes.

Within three months after the crash, the scar on the face of the Empire State Building had been repaired. Today the only trace of the tragedy is a small blackened crevice in the limestone blocks under the northwest windows on the 79th floor, seen only by window washers.

The Confederate Plot
To Burn New York

As the year 1864 was drawing to a close, the handwriting was on the wall. The Confederacy was not going to win the Civil War. But could a compromise somehow be made – one that would give the South the independence it sought in exchange for peace? Not likely, unless a wild scheme could force the North to the bargaining table. In desperation, Robert Cobb Kennedy, a Confederate Assistant Inspector General, who had been booted out of West Point before the war, devised a scheme he was convinced would work. He would burn New York!

Kennedy had been captured and was imprisoned at Johnson's Island in Lake Erie, from which he had escaped and fled to Toronto, Canada, a city full of Southern refugees and other escapees. He joined forces with some of them and devised an incredible scheme. They would travel to New York City, where some, led by Kennedy, would set a series of fires as a diversion. Others would seize Federal buildings and municipal offices, take control of the police department, free prisoners held at Fort Lafayette, and throw Union Army Commander Dix into a dungeon. By evening the Confederate flag would fly over New

An artist's rendition of one of the arsonists setting fire to Room 108 at the Tammany Hotel. Although the evilly leering face does not fit, it was Robert Cobb Kennedy who torched the room. From *Harper's Weekly*, December 17, 1864.

York, and negotiations could begin to end the war, with southern independence assured.

Eight of the conspirators were to travel to New York and use "Greek fire" for their incendiarism. This was a mixture of phosphorous in a bisulfide of carbon, which, upon contact with air, would ignite within minutes. The eight men, dressed in civilian clothes, boarded a train for New York, where a chemist with confederate leanings would prepare the Greek fire.

Each man, except Kennedy, stayed in a different hotel or boarding house. They moved each day or two. But Kennedy decided to remain in Mrs. Oatman's Boarding House for over two weeks, as he was having an affair with chambermaid Ann Cullen.

Kennedy and four other co-conspirators planned to torch at least two hotels each. They would check in, disperse Greek fire in their rooms, and leave quickly.

The Greek fire would take several minutes to produce the chemical reaction resulting in ignition.

One of the group went to the chemist's shop near Washington Square. Asking for his "luggage," he was handed a heavy valise, which contained 144 sealed bottles of a clear liquid. Now the schemers were ready to carry out their plot, so they set 8:00 p.m. on Friday, November 25, 1865, as the time for action.

Kennedy checked into room 58 at the New England Hotel, where he placed the bedclothes and mattress in a pile, put a chair on top to keep the pile steady, poured the contents of a Greek fire vial onto the pile, and walked out of the hotel. He walked ten blocks to

Robert Cobb Kennedy three days before his execution at Fort Lafayette in New York Harbor. He sent copies of this photograph to his family in Louisiana. (Courtesy of Mr. and Mrs. W. A. LaFleur.)

Lovejoy's Hotel, checked into room 121, and repeated the process. Meanwhile, his accomplices were doing the same thing at ten other hotels.

The first alarm sounded at 8:43 p.m. for the St. James Hotel, where room 85 was full of smoke. Within minutes an alarm came in for the St. Nicholas Hotel, where smoke billowed from room 124. The watchman at the City Hall fire tower was ringing the huge 23,000 pound bell one stroke, and then four strokes, indicating fires in both the first and fourth districts. The alarms were picked up by watchmen at other fire bell towers throughout the city. Soon there were eight strokes for a fire in the eighth district.

The bells rang non-stop as alarms kept coming in for hotel fires – LaFarge House, Metropolitan, Astor House, Howard, Fifth Avenue, Tammany, Lovejoy's, Belmont, United States, as well as an alarm for Barnum's Museum. It was pandemonium!

But, amazingly, not a single life was lost, and each fire did only minimal damage. Had the conspirators used the types of accelerants used by arsonists today, the story could have been entirely different, but the "Greek fire" was ineffective, providing lots of smoke, but little fire. Also, the New York firemen did an exemplary job that night. The rivalry and rowdiness often associated with New York volunteer firemen of that period were absent as they all pulled together to thwart the pending disaster to their city. Otherwise there could have been catastrophic losses. Weak Greek fire and dedicated firemen had averted what might have been a horrible disaster.

The next morning, General Dix issued the following proclamation:

A nefarious attempt was made last night to set fire to the principal hotels and other places of public resort in the city. If this attempt had succeeded, it would have resulted in a frightful sacrifice of property and life. The evidence of extensive combination, and other facts disclosed today, show it to have been the work of Rebel emissaries and agents. All such persons engaged in secret acts of hostility here can only be regarded as spies, subject to martial law, and to the penalty of death. If they are detected, they will be immediately brought before a court-martial or military commission, and, if convicted, they will be executed without the delay of a single day.

Detectives fanned out throughout the United States and Canada, and one by one the arsonists were rounded up. At Kennedy's trial, one of the witnesses testifying against him was Ann Cullen, the chambermaid, whom he had spurned. He was found guilty and sentenced to death by hanging. A wooden marker at the Soldier's Burying Ground outside Fort Hamilton reads:

R.C. Kennedy
Rebel Spy
Executed
Mar. 25, 1865

Fifteen days later, the South laid down their arms. The war was over.

The Civil War Fire Zouaves

Within three days after the outbreak of the Civil War in 1861, President Abraham Lincoln issued a desperate appeal for troops to enlist in the Union Army for a mere three months to "put down the rebellion."

A friend of Lincoln, young West Pointer Colonel Elmer E. Ellsworth, soon arrived in New York City for the purpose of recruiting fighting men. Although the city's volunteer firemen were exempt from military service, 1,100 of them, motivated by patriotism, rushed to sign up.

The dictionary defines "Zouave" as "a member of a French infantry originally wearing brilliant uniforms and conducting a quick spirited drill."

The dictionary defines "Zouave" as "a member of a French infantry originally wearing brilliant uniforms and conducting a quick spirited drill." The Fire Department outfitted the enlistees with bright red shirts, gray jackets, and flowing gray trousers tucked into their boots. Their colorful new uniforms, together with their fighting spirit, quickly earned them the title of "Zouaves," and they were a stirring sight as they marched

Zouaves are shown in this rendering battling the Willard Hotel Fire.
They saved the hotel.

The New York Fire Zouaves parade down New York streets wearing their colorful new uniforms.

through the streets of New York on their way to board the ship that would carry them to Washington, D.C., to be sworn in.

Arriving at the nation's capital, as President Lincoln looked on, they were mustered into service on the steps of the Capitol Building. The "Zouaves" were raring for action.

They soon found some, but it wasn't against the Confederate Army. They were actually billoted in the Capitol and slept on its floors. Soon they got word that Washington's most prestigious hotel, The Willard, was on fire, and the Zouaves raced to the scene, commandeering various pieces of Washington's fire fighting equipment along the way.

They fought the fire with such zeal and tenacity that they saved the hotel, which would have been doomed but for their efforts. The hotel owner, Mr. Willard, was so pleased that he served them all

Zouave leader Col. Ellsworth becomes the first Union officer to lose his life in the Civil War.

breakfast and wrote out a check for $500.00, no small sum in those days.

Two weeks later the Zouaves were ordered to march to nearby Confederate held Alexandria, Virginia, and occupy the city. As they marched into town, Col. Ellsworth spotted a Confederate flag flying atop the small Marshall House Hotel. Entering the hotel alone, he ascended to the attic and pulled down the flag. Suddenly confronted by the proprietor, Ellsworth exclaimed, "I claim this flag!" "Then, sir, I claim you!" came the proprietor's response as he shot Ellsworth dead. Ellsworth became the first Union officer to give his life in the War Between The States.

When the original Zouaves were mustered out after their three month enlistments were up, only 380 of the original 1,100 volunteer firemen were left. Sadly, the others had been killed, wounded, or captured at the Battle of Bull Run.

Additional regiments of Fire Zouaves, some from Boston, Philadelphia, and various other cities, entered the Union Army, and their heroism was notable. On the battlefield at Gettysburg still stands a statue of a Fire Zouave and a soldier, side by side.

Motorcycles That Fought Fires

It was a concept from the 1920's that really worked — motorcycles laden with firefighting gear that were able to weave in and out of traffic, get to the fire fast, and put it out while it was still small, sometimes minutes before larger apparatus arrived.

So successful were motorcycles in fighting fires, that after the Richmond, Virginia, Fire Department put their first one into service in 1920, they kept adding more, and by the end of 1922 they had purchased seven. A decade later, in 1931, they had nine motorcycle firefighters in service, and in 1938 they had a fleet of ten.

In the early 1920's, the Indian Motorcycle Company (at that time a formidable competitor to Harley-Davidson) was looking for ways to utilize their huge manufacturing capacity, so they teamed with the Foamite-Childs Corporation to produce the "Indian Fire Patrol."

The apparatus was really an Indian "Big Chief" 74 cu. in. motorcycle with a side car equipped with fire extinguishers,

An illustration from the Indian Fire Patrol sales brochure.

axe, pike pole, steel broom, siren, and perhaps most important of all – 100 feet of 3/4" garden hose with a hydrant adaptor and wrench. All the arriving firemen had to do was to connect the hose to a hydrant, turn it on, and squirt. Of course, the fire had to be conveniently within 100 feet of a hydrant. However, an optional model was introduced in 1927 which carried a 25 gallon soda-acid chemical tank.

According to one report, the Indian Fire Patrol was the hit of the 1923 International Fire Chiefs Convention, where it was introduced. The list price was $785.00, and the first production run of 50 sold out quickly, necessitating a much larger production run of 200.

Cities putting the Indian Fire Patrol into service in addition to Richmond, Virginia, included Boston, Massachusetts; Duluth, Minnesota; and Charleston,

South Carolina. Perhaps the greatest testimony to the efficiency of the Indian Fire Patrol was the letter from Chief L.S. Jones of the Richmond Fire Department, which said:

"We now have five of these fire-fighting machines in our Department, and hope to add more shortly. These little 'fire-bugs' have proven far more efficient than we ever dreamed they would. During 1923 they responded to 325 alarms, only two of which needed other apparatus to extinguish the fires."

With a record like that, firefighting motorcycles might well have been expected to sweep the country, but such was not to be. After American-LaFrance bought out Foamite Childs in 1927, production of the Indian Fire Patrol ceased. The last one was manufactured in 1928. Yet, for ten more years they still raced to fires in Richmond, Virginia.

This determined firefighter is ready to race his Indian Fire Patrol to the next alarm.

The Fire-Fighting Trolley Car

The Park Point area of Duluth, Minnesota, is a strip of land only 500 to 600 feet wide, reaching out six or seven miles to the Wisconsin shore of Lake Superior, and along that narrow strip are numerous summer cottages as well as expensive residences. Only one road goes down the strip – Minnesota Avenue – with a trolley track down its center.

How could the city of Duluth, in the days before the advent of motorized fire apparatus, provide fire protection to its citizens who lived along that narrow strip of land? Someone suggested that the trolley track held the answer. Why not? A trolley car fire engine went into service and worked like a charm, providing excellent fire protection to Park Point citizens for more than two decades – from 1907 to 1930.

For $650 the city of Duluth purchased an electric trolley car from the Interstate Traction Company, which also donated a car barn to house the new "fire engine" – the only one of its kind ever known to exist in America. The car barn became Duluth Fire Station No. 5.

Volunteer firemen removed the seats, and in their place installed a hose bed holding 1500 feet of 2-1/2"

The trolley car that responded to fires from Duluth, Minnesota Fire Station No. 5.

hose. The hose bed was open at both ends so that hose could be laid in either direction. Additional equipment included a pump of unknown capacity, a 35 gallon soda-acid chemical tank with 200 feet of 3/4" hose, ladders, axes, pike poles, extinguishers, and turnout gear.

When an alarm for Park Point tapped in, the trolley started down Minnesota Avenue, clanging its bell to alert the volunteers who lived along the track, and the trolley would stop momentarily to let them hop aboard.

When the fire-fighting trolley arrived at the fire with its compliment of volunteer firemen aboard, clad in their turnout gear, they would lay out their hose and attack the flames – a job they willingly and successfully performed for twenty-three years.

The Fire That
Made Beggars Rich

The *SS Fort Stikine*, an ammunition ship operated by the British Ministry of War Transport during World War II, arrived in Bombay, India, from England on April 14, 1944. Her cargo included 1,400 tons of explosives, war supplies, cotton, and – it was rumored – $6 million in solid gold bars.

The Bombay Fire Department consisted of 72 pieces of apparatus. Although most of the equipment was decades old, several pieces were relatively new 1000 gallon per minute pumpers. There were no fireboats to protect the ships at anchor nor the congested high-value area around the docks. Since the onset of the war, the department had successfully fought 60 shipboard fires, but Chief Coombs knew that if an ammunition ship ever caught fire and blew up, Bombay would be sprayed with flaming debris that would start a conflagration in the city of one and one-half million inhabitants.

What Chief Coombs feared came to pass. On board the *Stikine*, longshoreman Taqis noticed wisps of smoke curling up from between the bales of cotton on which he was standing. He yelled "Fire!" and his boss pulled a fire alarm station on the ship. Crewmen opened fire hoses onto the burning bales, but after 20

As the first Bombay Fire Department pumper pulled up, the fire in the hold was already blistering the paint on the side of the ship.

minutes they had made no progress, and the Bombay Fire Department still had not been called. Only after a city fire alarm box was finally pulled, more than an hour and a half after Taqi had discovered the fire, did the fire department respond.

> *Dozens of 280-pound bars of solid gold rained down upon Bombay. They landed in bazaars, gutters, rooftops, and crashed through windows.*

Some 32 hose lines were pouring water into the hold, but still the smoke continued to grow in intensity. Chief Coombs questioned the captain and was both flabbergasted and horrified to learn that directly under him were 370 tons of TNT. The captain also revealed that in the hold were 124 ingots of pure gold which the Bank of England was sending to the Reserve Bank of India to stabilize the rupee.

In addition to the 370 tons of TNT, there were over a thousand tons of other explosives on board. With the fire now blistering the paint on the *Stikine's* port side, Chief Coombs yelled, "Get off the ship! It's going to blow!"

Truer words were never spoken. The *Fort Stikine* exploded with such a thunderous boom that it not only rocked Bombay, but was recorded on a seismograph a thousand miles away at the Simla Meteorological Station in the foothills of the Himalaya Mountains.

Dozens of 280-pound bars of solid gold rained down upon Bombay. They landed in bazaars, gutters, rooftops, and crashed through windows. Beggars fell to their knees, scrambling for the gold that was falling from the sky. The lucky ones became rich.

A fire captain, holding a blackened slab of metal fifteen inches long, ran up to Chief Coombs. Coombs scraped away the soot and exclaimed, "It's a hunk of solid gold!"

Not until 48 hours later was the fire under control. Chief Coombs and his men saved Bombay, but at a terrible price. The number of buildings destroyed was 586. Of the 72 pieces of fire apparatus, only 37 were left. Of the 156 members of the Bombay Fire Department, 66 were dead or dying, with many others badly hurt.

No one has ever said how many gold bars were recovered, and after more than a half century, occasionally a gold ingot still turns up. Is there still more gold scattered or buried along Bombay's waterfront, or have now wealthy former beggars found it all? We'll likely never know.

The
Circus Fire Mystery

The Ringling Brothers and Barnum & Bailey Circus, termed "The Greatest Show On Earth," opened for their matinee performance on a humid summer afternoon in 1944 in Hartford, Connecticut. Covering more than one and one-half acres, the "Big Top" was big indeed, measuring 425 by 180 feet.

Over 6000 people, most of them children, crowded under the Big Top to watch lions and tigers, the world-famous Flying Wallendas, and Emmett Kelly, the circus clown known to millions. Thousands of enthralled persons watched the spectacular entertainment, never suspecting the canvas tent had been treated with a weatherproofing mixture made of highly flammable paraffin thinned with gasoline. It was a disaster waiting to happen.

Just as his performance was about to begin, Karl Wallenda of the famous trapeze act spotted a small fire in the grass under the bleachers and called to Emmett Kelly, who ran for a pail of water. Many in the audience supposed it was a circus stunt. But it was no stunt; it was a stunning tragedy in the making.

Suddenly flames shot up a rope to the top of the tent, then mushroomed to produce the most deadly circus disaster of all time. Children screamed as their parents grabbed them and dashed for the exits, but some of the exits were blocked by the chutes through which the lions and tigers entered the circus arena.

> *Either fatally burned or trampled to death were 168 people, 67 of them children.*

Aware of serious trouble, band director Merle Evans, struck up "The Stars And Stripes Forever," the circus signal for an emergency.

Before all of the audience could make it out of the flaming tent to safety, the tent collapsed. Either fatally burned or trampled to death were 168 people, 67 of them children. An Associated Press reporter wrote, "The scene of America's worst catastrophe was a day in which coffins became scarce and clowns cried."

The aftermath of the tragedy produced a mystery that went unsolved for forty-seven years. One of the children who perished was a little girl who had been untouched by the flames. She merely appeared to be peacefully asleep. Authorities were unable to identify her even though newspapers around the world carried her picture. She came to be known simply as "Little Miss 1565," which was the number on her tag at the morgue. For nearly half a century no one knew who she was.

The mystery of "Little Miss 1565" so intrigued Lt. Rick Davey of the Hartford Fire Marshal's Office that he spent many hours of his own time, over a period of years, chasing down clues, tips, and weighing evidence. In 1991, forty-seven years after the tragedy

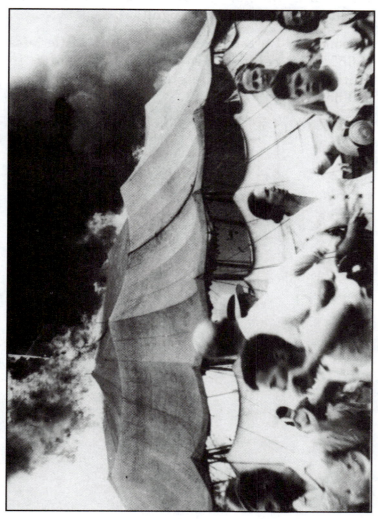

Survivors flee the Ringling Brothers and Barnum & Bailey big top as flames envelope the canvas which had been waterproofed with paraffin and gasoline.

had occurred, he positively identified the mystery child as nine-year-old Eleanor Emily Cook, who had attended the circus with her mother and older brother. Her brother also had perished. Although her mother had been badly burned, she lived. She was unaware that her daughter's picture had been in the newspaper and assumed that she had been burned beyond recognition.

Soon after Lt. Davey's identification, "Little Miss 1565" was reburied in her family's plot in Southhampton, Massachusetts – this time under her real name.

Colonial America's Foremost Fire Engine Builder - An Apprentice Of Paul Revere

Boston's volunteer fire companies during the 1700's included on their rolls of active firefighters great American leaders such as John Hancock, Samual Adams, John Adams, and master craftsman Paul Revere.

Paul Revere operated a foundry that employed a number of young apprentices including William C. Hunneman, who was destined to become America's foremost builder of hand fire engines during the Colonial Period and beyond.

After learning his trade as a coppersmith, brass founder and finisher, the young Hunneman, born in 1769, and now in his twenties, struck out on his own, turning out such mundane wares as pots, pans, kettles, and hardware for the citizens of Boston.

But when several major fires struck Boston and devastated several sections of the city, Hunneman resolved to improve on the bucket brigades and primitive hand engines of the day. He purchased the fire engine patents of inventor Jacob Perkins and configured those ideas into the celebrated Hunneman Fire Engine, which soon became famous for its reliability, durability, and superior drafting capability. So well

William C. Hunneman
1769 - 1856

engineered and constructed were Hunneman engines that they were in demand not only by fire departments throughout the United States, but throughout the world as well.

Hunneman manufactured his first engine in 1792, and during the next 90 years, he and his sons, who kept the business flourishing after his death, turned out more than 700 hand fire engines which were delivered to 26 states as well as to such diverse countries as China, Egypt, Cuba, Mexico, Haiti, Chili, Philippines, Turkey, and Australia. The Boston Fire Department alone had 30 Hunneman engines in service.

It took at least one month for the Hunneman Foundry to turn out a single engine. Although hard for us to imagine today, they were produced without the benefit of modern mass production methods. All

An early Hunneman engine, 1821, for Wrentham, Massachusetts. Note that the nozzle was almost five feet long.

engine parts were hand-forged, and while the brass and copper parts were rough-cast in the foundry, they had to be finished by hand. Wheelwrights had to shape rims and carve spokes. Hubs had to be turned, and cabinet makers formed trunnions. Woodworkers artistically carved rosettes, leaves, and other ornamental decorations ordered by fire companies that wanted their engines to be showpieces as well as firefighting machines.

The engines usually left the Boston factory beautifully lettered and decorated with scrollwork showing the name of the engine, and often the slogan of the fire company to which it was destined. Names describing the engines' capabilities included "Cataract," "Torrent," "Neptune," "Rapid," "Vigilant," "Alert," and "Water Witch." Slogans included "Our Duty Is Our Delight," "Douse The Glim," "We Will Try," and "Now With A Will Give Her Hell."

An example of the endurance of Hunneman engines is the "Papeete," shipped in 1852 bound for the island of Tahiti. It never arrived at its Pacific destination. When the ship docked at San Francisco to be reprovisioned for the second half of its voyage, the captain as well as the entire crew abandoned the ship to seek their fortunes in California at the height of the Gold Rush.

Soon a large fire struck San Francisco, and the "Papeete" was commandeered from the ship at anchor to help fight the flames. It rendered such noble service that it remained in California faithfully fighting fires for over one hundred years.

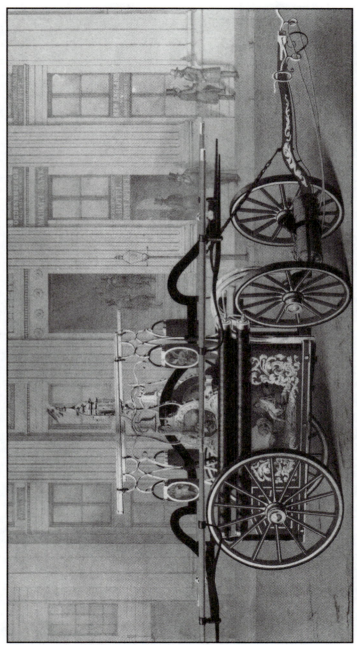

An ornate Hunneman engine manufactured in 1855 for St. Louis. Note the risque oil painting on the side of the tank.

The Fire Department
Without A Pumper

By hiring seven paid firemen to man seven brand new pieces of apparatus, a new fire department came into being in 1881. Strangely, not even one of the seven new pieces of equipment was a pumper, nor did the new fire department even purchase one during the next thirteen years.

Those were the days of hand pumpers in small departments, and steam pumpers in larger paid departments. But the Baltimore County (Maryland) Fire Department chose not to have any of them on the roster.

A decade earlier, "chemical engines," invented in France, made their appearance in America. They carried one or two 50 to 100 gallon tanks of water into which was mixed ordinary baking soda – the kind that's found on every kitchen shelf. A glass bottle at the top of the tank contained sulphuric acid. By tipping the bottle with a lever, the acid poured into the soda water to create an instant chemical reaction producing carbon dioxide gas within the tank reaching pressures of as much as 200 pounds per square inch, forcing the soda water out the hose and nozzle.

Charles T. Holloway, who was both a Fire Chief and a manufacturer of chemical fire engines.

With that kind of pressure, who needed a pump? Not the Baltimore County Fire Department, who resisted an engine with a pump for thirteen more years. On every alarm, all seven Holloway chemical engines responded from seven stations, each one pulled by two horses.

More and more Holloway chemical engines were added to the department's roster until they totalled 23, and still there was not a single pumper. Finally, in 1894, they broke down and bought two steam pumpers – plus six more Holloway chemical engines. Now the roster showed two steamers and 29 Holloway chemical engines.

During this period there were two principal manufacturers of chemical engines in the United States – one in Chicago, and one in Baltimore. The one in Baltimore was owned by Charles T. Holloway. And, by the way, the chief of the Baltimore County Fire Department was none other than Charles T. Holloway.

One of the 29 Holloway chemical engines in service in the Baltimore County Fire Department.

A Thing Of Terror

By the year 1872, steam had been in use for two decades to power the pumps on horse-drawn fire engines. Although early attempts to harness the steam's power to propel the engines to fires had been unsuccessful, inventor Neamiah Bean was determined to make it work.

An employee of the successful steam fire engine manufacturer Amoskeag, Bean worked tirelessly until he had created a monster weighing 17,000 pounds – the largest steam fire engine in the world. Rated as a "double-extra first size" engine, it could not only pump 1,300 gallons of water a minute, but could actually propel itself to fires without the use of horses at the incredible speed of ten miles per hour.

Bean worked tirelessly until he had created a monster weighing 17,000 pounds – the largest steam fire engine in the world.

Its success can be measured by the fact that after the huge engine had been sold to Bean's home town of

Striking terror into the hearts of persons in the way, this Amoskeag self-propelled steamer is shown speeding (at 10 miles per hour) to a fire in Boston.

Manchester, New Hampshire, for $3000, it remained in faithful service for the next seventeen years.

But Bean's huge self-propelled steam fire engines were not without their detractors. An issue of *The Fireman's Journal* ran the following story:

> A steam propelled fire engine is a thing of terror. It seems to us that a steam propelled fire engine is a terrific menace to life and property. The fire engines, in rushing through the streets, are dangerous, but the horses act as fender in case a person is unfortunate enough to get in the way. But with the steam propeller tearing down the street, it would be like an express train ploughing its way through the streets.
>
> Any person who has seen Jumbo No. 1 of the Hartford, Connecticut Fire Department – a steam propeller – knows what a thing of terror it is.

Jumbo has several names on its death roll. Its first driver met death in turning a corner, teams have been run into, and it was only a year ago that at an exhibition a horse and its rider met their death under its crushing wheels.

Although the 22 steam propellers of Bean's design, manufactured by Amoskeag, remained in service for many years, they never gained popularity and accounted for only a tiny fraction of all 5000 or so steam fire engines which were ever manufactured in America.

Yet, to Neamiah Bean's invention goes the credit for being the very first commercially produced, successful, self-propelled road vehicle in the United States.

The Men Who Wore These Helmets Weren't Firefighters

It was 2:00 a.m. when the loud gong pounded out the box number and shattered the sleep of the six men on the second floor of the station. Rising from their beds, they slipped into their boots, hitched their suspenders, and slid down the brass pole to the apparatus floor below. Within moments the apparatus motor roared to life, and the truck cleared the station door and accelerated down the street with its red lights flashing and siren screaming.

But the station was not a fire station, the apparatus was not a fire truck, and the six men responding to the fire alarm were not firefighters.

The crew, not members of the fire department nor paid with taxpayer dollars, were a company of the Fire Patrol or Salvage Corps, who were responding to the fire on a truck loaded not with hose or ladders, but with dozens of salvage covers. These canvas tarpaulins would be used to cover tens or even hundreds of thousands of dollars worth of goods and merchandise to protect it – not from the fire – but from thousands of gallons of water directed onto the flames by the fire department.

A horse-drawn Salvage Corps wagon circa 1899.

Organized and paid by fire insurance companies, the Fire Patrols or Salvage Corps operated in at least two dozen major American cities from 1869 to 1962, including New York, Chicago, Boston, Philadelphia, Baltimore, San Francisco, Indianapolis, and Louisville.

The equipment most often used by the patrols, in addition to salvage covers, included portable water shoots, piercing poles, sprinkler head kits, brooms, squeegees, sawdust, and portable pumps to drain basements.

An example of the typical work of the Patrol comes from the records of the Chicago Fire Insurance Patrol during the years 1896 through 1900, when they responded to 2,729 alarms and spread 7,646 salvage covers. The value of the merchandise saved, although not recorded, was many millions of dollars.

On Sunday, February 7, 1904 at 10:48 a.m., the Baltimore Fire Insurance Salvage Corps responded to

Over the nearly one century that fire patrols were active, many daring rescues by members were recorded, and dozens of members lost their lives in the performance of their duties.

an automatic alarm at the John W. Hurst Co. on German Street. When Captain Jordan arrived and saw smoke issuing from the sidewalk grating over the basement windows, he immediately drove his wagon to nearby Box 447 and personally struck the first alarm to the fire department for the Great Baltimore Fire.

Before it was controlled by the Baltimore and mutual aid fire departments responding from as far as 100 miles distant with steam fire engines hastily loaded onto railroad flatcars, the fire had consumed 1,526 buildings. Salvage Corps members were on continuous duty for 96

A 1915 American LaFrance Salvage Corps apparatus.

hours without sleep, and they guarded various damaged buildings for another 60 days.

Over the nearly one century that fire patrols were active, many daring rescues by members were recorded, and dozens of members lost their lives in the performance of their duties.

Today the work of the Salvage Corps is done by the fire departments themselves, but during the era of their existence, the now almost forgotten Fire Patrols were a vital adjunct to America's fire service. Measured in today's dollars, they saved their employers – fire insurance companies – claims totalling billions of dollars.

A 1930 Ahrens-Fox Salvage car for the city of its manufacturer - Cincinnati, Ohio.

Richard Henrich Collection

The Day Pennsylvania's State Capitol Burned

The Pennsylvania State Capitol building, built in 1822, was an imposing structure 180 feet long and 80 feet deep, with its majestic dome rising 130 feet into the air over the city of Harrisburg.

Evidently little thought had been given to fire protection, even by 1897, when the venerable capitol was 75 years old, as evidenced by the fact that none of the fire hydrants on the capitol grounds had threads compatible with the hose in Harrisburg's eleven fire companies.

Moreover, any competent fire inspector would have noted that the frame construction of the building's interior was completely dried out, that there were heavy coats of paint and varnish on the woodwork, that there were heavy accumulations of dust beneath the flooring as well as in the hot air ducts, and that the rotunda beneath the dome would act like a huge flue if ever it caught fire.

About 10:30 a.m. on February 2, 1897, Senator Grady thought he detected the slight odor of smoke and sent a page boy to investigate, who returned to report nothing amiss. About 11:15 a.m. Senator Grady was sure he smelled smoke, and this time

The Pennsylvania State Capitol as it appeared before the fire in 1897.

he sent a different page boy to investigate. This boy found nothing either.

Finally the smell of smoke became so strong that several senators set out on their own to investigate, and they discovered smoke seeping from under the door of the Lieutenant Governor's Office. Once inside, as they used a hatchet to chop through the flooring, flames burst forth. "Get your buckets!" shouted Chief Clerk Edwin Smiley to the group of senators who had gathered at the fire scene, but no thought was given to alerting the fire department.

Finally Librarian Miller pulled an auxiliary fire alarm connected to the master fire alarm box No. 2-3-2 at the corner of Fourth and State Streets. But for whatever reason, the alarm tapped in not as 2-3-2 but simply as 2-3, and the fire department, including Harrisburg's only ladder truck, responded to the Pennsylvania Railroad Shops instead of the State Capitol building.

Long hose lays were required because the capitol's hydrants had the wrong threads with no adaptors, and the ladder truck was at the wrong location, so it soon became obvious that the capitol was doomed.

"The Pennsylvania State House Fire" by artist Landis Brent Whitsel.

(Copies of this print in full color are available from The Fire Museum of Greater Harrisburg.)

Every piece of Harrisburg fire apparatus, including seven steamers (two of which were 29 and 33 years old), and every foot of fire hose in the city could not stop the now roaring inferno.

Finding their way up the rotunda to the historic dome, flames broke through around the dials of the tower clock, which tolled the hour of one o'clock. It was never to strike again. Soon the dome collapsed with a roar some thought sounded like thunder, as men, women and children watched, many with tears in their eyes.

By 4:00 p.m., the proud State Capitol building, the pride of Pennsylvania, had been reduced to rubble and ashes. Scarcely had the ashes cooled when the city of Philadelphia mounted a vigorous campaign to move the state capital from Harrisburg to "The City Of Brotherly Love." Newspaper articles were quick to point out that Philadelphia had a superior fire department.

But the Harrisburg location prevailed, and in 1906, a brand new capitol building was completed on the exact site of the ill-fated previous structure. Harrisburg residents were doubly proud of their new capitol building when it was dedicated by Theodore Roosevelt, the President of the United States.

The Suspicious Cruise Ship Fire
That Drew One Million Spectators

The fast, luxury cruise ship *Morro Castle* was returning to New York City on September 8, 1934, from Cuba, where her 555 passengers and crew had been shopping and partying in Havana.

Built just four years earlier in Newport News, Virginia, the *Morro Castle* was a twin screw turbine electric liner 508 feet long with a beam of 70 feet, and she featured the latest state-of-the-art safety features. There were nine watertight bulkheads with electrically operated watertight doors, automatic smoke and fire detection systems, fire alarm signal stations throughout the ship, 42 fire hose stations, two fire pumps, and 12 lifeboats.

But the designer of all of these safety features had not reckoned with George Rogers, the ship's radio operator, who allegedly murdered the Captain and then set the ship on fire to destroy the evidence. At first hailed as a hero for sending an S.O.S. signal, Rogers was later suspected of poisoning the Captain and then setting the blaze with a delayed fuse device to destroy the evidence.

As the *Morro Castle* passed the Barnegat Light about midnight, most of the passengers and crew were asleep in their cabins. But Captain Wilmott did not wake up when the fire alarm sounded. He was dead.

The fire was discovered in the writing room on the port side of the promenade deck. The crew kept trying to control the fire with extinguishers but to no avail. Hours before his suspicious demise, Captain Wilmott had turned off the smoke detection system to avoid its activation by the fumes of salted hides which he was transporting from Cuba.

But as the fire continued to grow in intensity, the alarm system was activated manually, which resulted in panic. Taking over for the dead Captain was William Warns, who chose to speed up the ship for early arrival in New York City, where the world's largest fire department could take charge and extinguish the fire. Meanwhile, the Chief Engineer, Eban Abbott, who should have directed the firefighting operations, remained in the deckhouse, frozen by fear.

As the flames continued to involve the center of the ship, rolling up elevator shafts, the *Morro Castle* lost power. She would never make it to New York. As the crew themselves began to scramble into lifeboats, the passengers huddled at the bow and stern of the ship. Unbelievably, Acting Captain Warms still refused to send an S.O.S.!

Finally, about 2:30 a.m., a passing ship, seeing the *Morro Castle* in flames, radioed the Coast Guard Station at Manasquan, New Jersey, who replied that they had received no distress call. Not until 3:15 p.m. did radio operator George Rogers send out an S.O.S.

By the time three ships in the area responded, 134

By the time three ships in the area responded, 134 of the 555 passengers and crew had perished.

of the 555 passengers and crew had perished. Many were trapped in their rooms and were suffocated or burned to death. Others jumped into the ocean with hard life rings around their heads and died of broken necks. Still others succumbed to hypothermia in the sea.

Of the 12 lifeboats, only 6 were launched, with the others burned at their mounts. With space for 400, the lifeboats carried only 85 survivors, most of them members of the crew. Chief Engineer Eban Abbot alighted from a lifeboat in his white dress uniform with nary a smudge.

Around noon the next day, Acting Captain Warms and the remaining survivors were removed from the stricken ship by the U.S. Coast Guard, but efforts to tow the still burning ship failed as the tow lines repeatedly burned through. The *Morro Castle* was drifting toward shore on her own, and at Asbury Park, New Jersey, Fire Chief Taggart watched with horror as the burning vessel appeared to be headed for the new Convention Hall, which extended 200 feet into the Atlantic Ocean.

But she ran aground short of the Hall, and Chief Taggart positioned two 1000 gpm pumpers to supply hoselines to fight the fire. Gradually Asbury Park firefighters extinguished the burning ship and recovered the bodies. The firefighting operation lasted for seven days. One of the two pumpers, a 1929 American LaFrance, restored in 1976, remains on the Asbury Park Fire Department roster to this day.

Courtesy of the News. New York's picture newspaper.

The burning *Morro Castle* drifted ashore within several hundred feet of Asbury Park's new Convention Hall. Over the next few days a million people watched the spectacle from the boardwalk and the beach.

George Rogers

As the *Morro Castle* billowed smoke onto the Asbury Park waterfront, hundreds of thousands of curious sightseers thronged to the area, and boardwalk concessionaires reopened to sell souvenirs from their shops which had just been closed for the season. The total number of spectators was estimated at over one million!

But what about radio operator George Rogers, the "hero" who finally sent the S.O.S.? His fate is bizarre. Investigators could never gather enough incriminating evidence to charge him with either murder or arson. But several years later while he was working at the Bayonne, New Jersey, Police Department, he boasted to his supervisor, Captain Doyle, exactly how he had constructed the delayed fuse device he had used to start the fire.

Captain Doyle immediately reopened the investigation, whereupon Rogers constructed yet another explosive device to blow up Doyle. But Doyle, although losing three fingers, survived the blast, and Rogers was charged with the attempted murder of a police officer and sent to prison. But the story doesn't end there.

During World War II, Rogers was somehow paroled in order to serve the war effort as a radio operator. After the war, he got into an argument with the

husband and wife who were his landlords and murdered them. He was convicted of their murder and
returned to prison, where he died of a heart attack in
1958.

Yet, Rogers was never charged nor convicted of
either murdering Captain Wilmott or setting the fire
that killed 134 persons aboard the ill-fated *Morro
Castle*.

The *Hindenburg* Disaster
And The Little Engine That Couldn't

No fire engine in existence either then or now, no matter how big or powerful, could have saved the *Hindenburg* airship, much less could the little 1934 Hale/Ford V-8 pumper from Mantaloking, New Jersey, that made a 15 mile run on a mutual aid call to help the doomed dirigible.

The huge lighter-than-air vessel designed and built by Count Von Zeppelin of Germany, was completed in 1936 after four years of construction. Eight hundred and four feet long, and 130 feet at her widest point, the *Hindenberg* was the largest man-made object ever to fly, even to this day.

Powered by six diesel engines, she had a two story passenger cabin which included a lounge with a piano, a kitchen, and a restaurant. She was the grandest airship ever built up to that time.

Count Von Zeppelin proposed to name the monstrous and elegant craft the *Adolph Hitler* and to inflate it with non-flammable helium gas. But he ran into stumbling blocks on both counts. First, since Hitler feared airships and refused to fly in them, the giant dirigible instead was named the *Hindenburg* in honor of the recently deceased presi-

Wide World Photos

German zeppelin, Hindenburg aflame as it attempted to land at the naval air station in Lakehurst, NJ in 1937.

dent of Germany, who was actually Hitler's predecessor. Second, because the United States was at odds with Germany and controlled the world's supply of helium, only the more volatile and flammable hydrogen gas was available to supply the lifting power to float the huge aircraft.

The U.S. Navy had a massive hanger at Lakehurst, New Jersey (it's still there today) that was the terminal for the Hindenburg on her round trips to Germany. Although she had already made 37 successful Atlantic crossings, crowds of onlookers still gathered at Lakehurst to watch her come in.

A ground crew of 250 men were needed to grasp the landing lines and guide the ship to its mooring tower, and they were ready for the ship's landing at 7:23 p.m. on May 6, 1937. But as they moved in to take hold of the lines, they saw a small flame atop the rear of the craft, and within moments there was a whoosh and popping noise as a giant fireball erupted. Within only

34 seconds the giant ship cracked and fell stern first to the ground as the fireball consumed seven million cubic feet of hydrogen in a spectacular explosion.

Mutual aid engines and rescue squads responded from as far as 30 miles away.

Amazingly, many of the 96 persons aboard the doomed *Hindenburg* survived by simply riding her down and walking away. But 13 passengers, 22 crew members, and one person on the ground perished. The cause of the fire and explosion was never definitely established, but theories included static electricity igniting venting hydrogen, as well as rifle fire from a saboteur.

The first fire apparatus to arrive on the scene was the 1934 Pirsch/Dodge of the Lakehurst Fire Department. Mutual aid engines and rescue squads responded from as far as 30 miles away.

But the little Ford pumper that made the 15 mile mutual aid run from the small oceanfront community of Mantaloking, New Jersey, is still maintained by Mantaloking Fire Co. No. 1, where it proudly displays its dash plaque presented by the United States Navy in appreciation for its response to the spectacular though sad moment in history when the giant *Hindenberg* perished.

Yesterday's Triggerless Dry Powder Extinguishers

By simply pulling the trigger, literally millions of modern dry powder fire extinguishers are ready at any moment to snuff out perhaps hundreds of small fires each day. They're under pressure, of course, and pulling the trigger sends the fire-quenching powder spewing out up to fifteen feet to reach the flames.

No so with the "tin tube" dry powder extinguishers manufactured from about 1900 to 1937. They required the user to forcefully swing the twenty-two-inch long, two-inch diameter tube back and forth, manually propelling the contents with centrifugal force.

Sold for nearly four decades by door to door salesmen for as little as $3.00, the tin tubes hung from hooks in hundreds of thousands of homes throughout the United States. Printed on the colorful tubes were instructions for their use, such as:

JERK OFF HOOK

THROW CONTENTS FORCIBLY

WITH SWEEPING MOTION

AT BASE OF FLAMES

The contents were usually just ordinary baking soda. The formula for the "Phoenix," one of the more popular brands, included phosphate of ammonia, white salt, pipe clay, yellow ochre, and Prince's metallic brown, all of which were added to the soda, but which added little, if any, efficiency to the extinguisher.

Although there were well over 200 different brands of tin tube extinguishers, virtually all of them were manufactured by one man, Edwin Norton, of Toledo, Ohio, who founded both the American and Continental Can Companies.

Norton "private labeled" the tin tubes during the manufacturing process while they were still flat, decorating them to his customers' specifications, including fancy designs and names such as **"Always Ready," "Blaze Killer," "Kilfyre,"** and **"Nevermys."**

Local companies bought the empty tubes and filled them with their own mix of baking soda and additives, then sold them door to door.

Today the old tin tube extinguishers often show up at flea markets, where they usually sell for $25 to $35, but some unusual examples in excellent condition can go for as much as $100. Often the contents are still intact, still ready to be jerked off the hook and thrown forcefully onto a fire with a sweeping motion.

Breaking The
Fire Lantern Color Code

Prior to the advent of electric hand lights about 1920, all fire apparatus was equipped with as many as six colorful kerosene lanterns. In fact, throughout the 1930's and even into the 1940's some fire departments, including New York City, were still ordering kerosene lanterns on new equipment.

The glass lantern globes not only came in solid colors such as red, blue, green and yellow, but in split colors, with the bottom half of the globe being clear, and the upper half colored. The purpose was not just to be fancy or to look pretty – each color or half color had a story to tell.

Let's move the clock back a century or so and watch the brilliantly colored lanterns moving about at the scene of a nighttime fire. If you know their long-forgotten code, you can understand what the different colors are saying:

Solid Red
Hose Company

Solid Blue
Engine Company

"KING" FIRE DEP'T LANTERN

REG. U. S. PAT. OFFICE—PATENTED

(HOT BLAST)

D I E T Z
"KING" FIRE DEP'T
LANTERN

D I E T Z
"KING" FIRE DEP'T
LANTERN
WITH GUARD RAISED
(BRASS)

Solid Green
Hook & Ladder Company

Split Red/Clear
Chief Engineer (Fire Chief)

Split Blue/Clear
Assistant Chief

Split Green/Clear
Foreman Hook & Ladder Co.

Split Yellow/Clear
Chief Hoseman or Assistant Foreman

Still another lantern – solid yellow – might sometimes be seen, and its owner was often considered the most important of all because he was the man with the money. The yellow lantern (sometimes referred to as 'gold') was held by the fire company's treasurer.

Fire Marks
The signs that sometimes told fire brigades to fight - or not to fight - your fire

In Colonial America, it was common to find a small metal sign, such as the one pictured below, affixed to the front of many houses and places of business. Adapted from use in Europe, early American fire insurance companies issued them to policyholders to mount on the front of their homes and buildings to indicate they were insured.

The first Colonial fire brigades were often sponsored by fire insurance companies. Benjamin Franklin founded the first company and brigade in 1752, the "Philadelphia Contributorship Fire Insurance Of Houses From Loss Of Fire." Their mark depicted four hands clasped and crossed, known as "Hand-In-Hand." But Ben, whose kite flying experiments had convinced him that lightning often caused fires, refused to insure homes near trees, which he believed attracted lightning.

Soon a competing fire insurance company sprang up to capitalize on Franklin's tree restriction. They advertised their willingness to insure dwellings regardless of how many trees were nearby. The fire mark of the new "Mutual Assurance Company" was, of course, a painted "Green Tree."

Firefighting brigades sponsored by Hand-In-Hand, Green Tree, and still other new fire insurance companies, often had their buckets and hand engines painted in different colors. When a fire alarm was sounded – usually by church bells – all the brigades turned out, but all of them didn't fight the fire. Incredibly, only the brigade sponsored by the fire insurance company which held the policy on the dwelling went to work, the others either staying to watch or returning to their quarters.

A St. Louis newspaper, in an article recounting that city's history of fire marks, stated the following:

"The idea was that you took out fire insurance on your house or store with a company which maintained its own firefighters. If and when the premises caught fire, you sent a messenger off at the run to the insurance company. The first thing the firemen did on arrival was to look for the metal fire mark, firmly affixed to your wall, as evidence that you were, in fact, insured with

Today, thankfully, virtually all fire departments assist each other whenever called upon to do so...

their company. If it wasn't there, they were likely to take their engine home again and let the place burn."

Today, thankfully, virtually all fire departments assist each other whenever called upon to do so, and it is unthinkable for a call for help to go unheeded, although a few isolated instances make the news from time to time.

But whatever became of the old firemarks of yesteryear? The few original ones that remain are eagerly sought by collectors, including the 350 members of the "Fire Mark Circle" – a group of fire mark enthusiasts who conduct a fire mark auction at their annual convention. The rarest fire marks sometimes go for more than $5000, but a word of caution – beware of reproductions.

Ben Franklin's "Ounce Of Prevention" Led To Volunteer Firefighting In America

"An ounce of prevention is worth a pound of cure" is one of the most famous and oft-repeated maxims in America. Yet few persons realize that it is attributed to Benjamin Franklin, who was a fire prevention and fire protection activist.

Here is the complete text of Franklin's famous quotation, which he penned in 1735:

"In the first place, as an ounce of prevention is worth a pound of cure, I would advise how they suffer living brands-ends or coals in a full shovel to be carried out of one room into another or up or down stairs, unless in a warmingpan and shut; for scraps of fire may fall into chinks and make no appearance until midnight; when your stairs being in flames, you may be forced (as I once was) to leap out of your windows and hazard your necks to avoid being over-roasted.

If chimneys were more frequently and more carefully clean'd some fires might thereby be prevented. I have known foul chimneys to burn furiously a few days after they are swept; people in confidence that they are clean, making large fires. Everybody among us is allow'd to sweep

Ben Franklin's Union Fire Company, organized in Philadelphia in 1736, is credited with popularizing the concept of volunteer firefighting in America.

A stylized portrait of Ben Franklin. The shed housing his Union Fire Company engine is in the background.

chimneys that please to undertake that business; and if a chimney fires thro' fault of the sweeper, the owner pays the fine and the sweeper goes free. This thing is not right.

Those who undertake the sweeping of chimneys and employ servants for that purpose, ought to be licensed by the Mayor; and if any chimney fires and flames out 15 days after sweeping, the fine should be paid by the sweeper; for it is his fault."

According to the book "Frankin and Fire" published in 1906, "One result of this paper seems to have been the founding of the Union Fire Company in 1736 by Franklin and four of his friends." They formed their fire company "for preserving our own and our fellow citizens' houses, goods, and effects in case of fire."

Benjamin Franklin and his Union Fire Company are credited by firefighting historians with popularizing the concept of organized volunteer firefighting in America.

Did Saboteurs Really Torch The Luxury Liner *Normandie?*

The United States had been at war only two months on February 9, 1942, when German propagandists heralded their first major sabotage success in America. The world's largest luxury liner, *Normandie*, was destroyed by fire while at anchor at New York City pier 88. She was in the process of being converted into a troopship, and Germany was taking credit for its demise.

With the recent Pearl Harbor attack fresh in their minds, Americans were fearful not only of sabotage but of an enemy attack or even an invasion. Air raid drills were held frequently. How could the *Normandie*, docked at the city with the largest fire department in the world, and with hundreds of workers on board, possibly be consumed by fire if it were not sabotage? The sabotage story was widely accepted until the National Fire Protection Association investigated and came up with the real facts.

The cause was traced to a welder's torch igniting a kapok-filled life preserver. Yet, how could a burning life preserver lead to the total destruction of the world's largest liner?

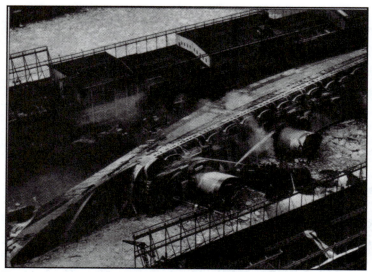

The world's largest luxury liner *Normandie* capsized after being destroyed by fire, which was first thought to have been set by saboteurs. But an investigation by the National Fire Protection Association determined the true cause.

The answer was because not one of the hundreds of workers or their supervisors on board the huge vessel thought to call the fire department! New York City firemen remained in their stations blissfully unaware that the *Normandie* was aflame. The NFPA termed the debacle a "Hollywood slapstick comedy."

By the time Box 852 was finally struck with the first of five alarms, the ship was doomed. Hundreds of firemen with land-based apparatus, as well as fire boats pouring millions of gallons of water, could not save the mighty liner, which capsized the following day.

America's war effort had lost its largest troopship, not from sabotage, but simply because no one called the world's largest fire department until it was too late.

Japanese Incendiary Bombs That Fell On The United States

No, the title of this story isn't mixed up, because incendiary bombs from Japan actually did fall on the United States during World War II, killing one adult and four children who were on a church-sponsored outing in the mountains.

During November 1944, 700 hydrogen-filled, bomb-laden Windship Weapons were launched, followed by 1200 in December, and 2500 during February and March of 1945.

The bombs were not carried from Japan in airplanes but were borne on the winds of the jet stream – that lofty river of air that blows from Japan toward North America in a great double arc that swings north to east, and then southeast over the United States.

The U.S. knew little about the jet stream in those days until our B-29 bomber crews encountered it late in the war when they flew with their bomb loads (including millions of incendiaries) high over Japan and found winds blowing up to 200 miles per hour. But the Japanese were well aware of the jet stream,

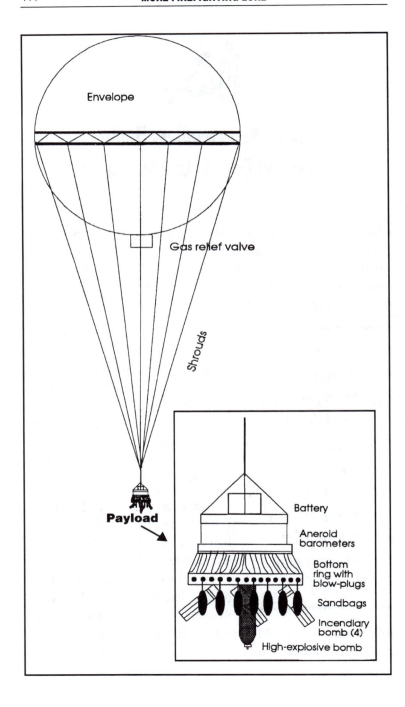

and they conceived a transpacific "Windship Weapon" to bomb the United States.

By experimenting with rubberized silk balloons that had radio-position altitude instruments on board, they established that balloons could reach North America carrying both explosive and incendiary bombs after climbing into the jet stream at 30,000 feet.

During November 1944, 700 hydrogen-filled, bomb-laden Windship Weapons were launched, followed by 1200 in December, and 2500 during February and March of 1945. The last 400 went up in April, when the program was terminated as being unsuccessful.

The Japanese military general staff had planned on the incendiary bombs starting forest fires in the Pacific northwest, while the explosive bombs would terrorize the population. A few fires were started, but no property damage resulted. None of the explosive bombs detonated, as far as is known, until pastor Archie Mitchell and his wife Elsie led five children ages 11 through 14 from their church in Bly, Oregon, on a Saturday morning fishing trip onto nearby Gearhart Mountain.

As they made their way up the south flank of the mountain, Elsie Mitchell stumbled upon something she did not recognize. "Look what I found!" she called out to her husband, but before he could get to her side an explosion rent the air. Pastor Mitchell found not only his wife, but all five of the children dead – killed by the only Windship Weapon launched in Japan and carried to America on the jet stream that performed its intended mission.

Unusual Locations
For Fire Stations

The Underwater Fire Station

With the pressing need for a third fire station to protect fast-growing Dublin, Ohio, with available land costing as much as $150,000 an acre, and with residents fussy about what went up near their homes, the Washington Township Fire Department had a problem. Yet, without realizing it, they were looking up at the solution every day.

> *A modern fire station was constructed under the tank – under two million gallons of water.*

The new 150-foot-high, two-million-gallon Dublin municipal water tank was already owned by the city. Why not build a fire station inside its supporting walls – beneath the water? The city agreed to lease the space for $1.00 a year, and the rest is history.

A modern fire station was constructed under the tank – under two million gallons of water. With apparatus at the ready on the ground floor, the second story includes a dormitory for the firefighters, along with restrooms, showers, stove, refrigerator, and pantry.

photo courtesy of Larry Eberhart

Dublin, Ohio's new fire station that just happens to be located under two million gallons of water.

But there's no traditional brass pole for the firefighters to slide down when an alarm hits. Instead, beginning with a nearly vertical drop and then easing off into a series of curves, a 44-foot-long plastic chute, made by a playground equipment manufacturer, deposits them, after a wild ride, within steps of their waiting equipment.

The Fire Station Inside A Mountain

No less innovative than the folks in Dublin, Ohio, are the firefighters in mountainous Creede, Colorado.

Their fire station is unique in the world, having been hewn out of the solid rock of a mountain.

Photo by R.W. Fortune

The Grand Masters Of Vintage Fire Engine Restoration

Throughout America are hundreds, if indeed not thousands, of full time restorers of vintage automobiles, but how many full times restorers are there of antique fire apparatus? Just two, and they each have enough work waiting for their unique expertise to keep them busy for several years into the future.

Ken Soderbeck of Jackson, Michigan, and Andy Swift of Hope, Maine, have several things in common. In addition to being the only two persons in the United States who make their full time living restoring antique fire apparatus, they have such a love for their work that they could not be paid enough to change careers.

And they have something else in common as well – even though their shops are a thousand miles apart, they're close friends, visiting each other once or twice a year and comparing notes over the phone at least weekly.

Collectors of vintage fire equipment soon learn that for an absolutely stunning restoration, whether for motorized, horse-drawn, or hand-drawn antique fire apparatus, Ken and Andy are tops in their field.

Andy Swift, owner of Firefly Restorations, was chosen to restore three of the engines in the celebrated Tour de LaFrance.

Collectors of vintage fire equipment soon learn that for an absolutely stunning restoration, whether for motorized, horse-drawn, or hand-drawn antique fire apparatus, Ken and Andy are tops in their field. With low overhead, their rates are reasonable, although customers may be dismayed when they learn that restoration on their treasured engine may not begin in weeks or months but perhaps in years.

Ken Soderbeck became a sign painter's apprentice in the early 1950's; then, he operated his own sign shop for 20 years. He learned the fine art of gold leaf decoration and is considered to be the ultimate master of this art, as it pertains to the ornate lettering and scrollwork on vintage fire apparatus.

But the decoration is only the final part of a total restoration project, which often begins by dismantling the engine, then sandblasting the parts down to bare metal in his quaint shop in a 100-year-old two-story schoolhouse, where he lives with his wife Agnes.

He recreates from scratch any missing parts such as engine lamps and upholstery. It all takes time. "I usually don't get a truck in and out of here in less than four or five years." But it's worth the wait, according to customers such as the American-

Ken Soderbeck is considered to be the ultimate master of gold leaf decoration as it pertains to the ornate lettering and scrollwork on vintage fire apparatus.

LaFrance Museum in Cleveland, North Carolina, and The Venerable Fire Collection, Inc. of Slinger, Wisconsin.

Similar work goes on up in Maine, where genial Andy Swift sometimes skips meals rather than interrupt his work on an old fire engine.

In 1977, he and his wife Kathy traveled to Valdez, Alaska, where he joined the fire department. During his off-duty hours he completely restored the department's 1907 steam fire engine. He was hooked. Now he'd rather restore old fire engines than eat.

Andy once had befriended none other than Ken Soderbeck, and with Ken booked years in advance, and with American-LaFrance needing two trucks restored as soon as possible, Ken asked Andy if he'd be interested in the job.

Would he ever! Andy jumped at the chance, moved back to Maine and turned out two restorations so stunning that American-LaFrance asked him to do several more, as have many museums and private owners during the past 15 years.

"Like most kids, I loved fire engines," says Andy, "and I guess that's one thing I just never outgrew."

The Kitty Litter Fire
That Extinguished Itself

If you tried your best and used your wildest imagination to combine the following ordinary items into a bizarre fire that put itself out, you probably couldn't come up with anything as strange as a 1997 fire in Brunswick, Maine. Here are the ingredients: (1) a litter box containing Fresh Scoop kitty litter (2) a bottle of hair gel (3) a bottle of aftershave lotion (4) a toilet (5) a cat.

Baffled? So were the Brunswick firefighters who responded to the alarm and found a bathroom which had obviously been on fire, no apparent cause of the fire, and no apparent reason for the fire to have gone out all by itself.

Yet, detective work by the apartment owner, the fire chief, and the Maine Fire Marshal's Office pieced together this incredible chain of events:

After the apartment occupant had left for the day, his cat jumped from the litter box on the bathroom floor onto the vanity, knocking over a bottle of hair gel so that it dripped into the litter. The result, according to investigators, who duplicated the result in the laboratory, was a spontaneous chemical reaction that ignited the litter into smoldering flames.

The fire spread to the floor, up a curtain, and onto the wall next to a bottle of aftershave lotion. The bottle burst, spraying the lotion containing alcohol over the wall, which also broke out in flames.

Next, intense heat shattered the ceramic toilet, which poured five gallons of water onto the floor to extinguish the fire there, and then the toilet's fill pipe sprayed more water onto the wall to extinguish the rest of the fire.

Perhaps the moral of the story is to keep aftershave and hair gel in a tightly closed container if you have kitty litter in your bathroom.

Fire Service Traditions

St. Florian - Patron Saint Of Firefighters

Two mysterious fires in the imperial palace in Rome in the year 304 A.D. were attributed to Christians by Emperor Diocletian, who issued an edict that all believers in Christ were to be slaughtered. But a 54-year-old high-ranking officer in the Roman Army named Florian, who was stationed in eastern Bavaria, refused to carry out the emperor's orders.

Learning of Florian's refusal to have the Christians in his area put to death, Emperor Diocletian dispatched his hatchet man Aquilinus to confront Florian. Aquilinus was even authorized to offer Florian a promotion if he would kill the Christians in his area, renounce his Christian beliefs, and worship the Roman god Jupiter.

"Go back to Diocletian," Florian replied, "and tell him that I am a Christian and will suffer all pain for Christ." Out-

St. Florian

Florian Hall in Boston, Massachusetts, is the headquarters of Local 718, International Association of Firefighters, and is an interdenominational facility often used for weddings and banquets. On the front of the building, on either side of the steamer mural, are the names of Boston firefighters who have died in the line of duty.

raged, Aquilinus sentenced Florian to death on the spot, and ordered his soldiers to build a funeral pyre.

Standing atop the pyre, Florian challenged the soldiers to light it, saying, "If you do, I will climb to heaven on the flames." Struck with fear and apprehension, Aquilinus considered Florian to be some sort of a magician, and ordered him removed from the pyre. But Florian's safety was short-lived. He was not to die by fire, Aquilinus decided, but by drowning.

Aquilinus had a millstone tied around Florian's neck, and had him cast into the nearby River Enns. Before hitting the water, Florian cried aloud, "Lord Jesus, receive my soul."

As he was martyred, it was reported that an eagle, with wings outstretched, flew overhead, casting a shadow in the form of a cross over the site of his drowning and kept watch until his body was retrieved by a pious matron named Valerie, who had his body carefully placed in a cart and taken to her country estate near Linz, Austria, where she had him buried in a tomb.

Nearly 600 years later, sometime between 900 and 955 A.D., a monastery was erected near the tomb, and the village of St. Florian grew up around it. One of the oldest monasteries in the world, it also is a sanctuary for art treasures.

Pilgrims assembled at the monastery's church each year on May 4, St. Florian's Feast Day, and participated in special services.

Made a saint by the Pope, St. Florian has been invoked against fire and has generally been regarded in most countries, including the United States, as the patron saint of the fire service. An excerpt from "Lives of Saints" by Alban Butler states, "He is invoked as a powerful protector to those in danger from fire."

The Origin Of The Maltese Cross

The story of the Maltese Cross, the firefighter's badge of honor, is nearly a thousand years old. When the Knights of St. John, a courageous band of Crusaders, fought the Saracens for possession of the Holy Land, they encountered a new and terrible weapon they had not known existed. It was a horrible weapon that brought them excruciating pain and an agonizing death. It was a fire bomb.

As the Knights of St. John advanced on the walls of the city, they were struck by glass bombs filled with naptha. The bombs broke and saturated them with the flammable liquid. Then the Saracens hurled a flaming tree into their midst.

As hundreds of Knights were burned alive, others risked their lives to save their brothers in arms from fiery deaths. Their valiant efforts were recognized by fellow Crusaders who awarded each hero a badge in the configuration of the cross which many firefighters wear today.

Since, for nearly four centuries, the Knights of St. John lived on the Mediterranean island of Malta, their crosses for heroism in rescue from fire came to be known as the Maltese Cross.

The Maltese Cross, as worn by firefighters today, symbolizes their willingness to lay down their lives for their fellow man just as the Crusaders did so long ago.

Sparky The Fire Dog

Sparky, the friendly-looking canine with the face of a Dalmation, dresses from head to tail in firefighter's gear and has been a familiar face to children for nearly half a century.

Sparky came into being in 1954 to carry the fire safety messages of the National Fire Protection Association, and he's been doing it ever since. Because Sparky has become a symbol of fire safety everywhere, his loyal following of millions of children is worldwide.

Sparky's Dalmation breed was an obvious choice for a fire prevention campaign since Dalmations were associated with horses in England for centuries, and then became firehouse companions to the horses which pulled fire apparatus to fires in America for a half century, beginning around 1860. Some Dalmations even rode on the apparatus to fires and were as much a part of the fire scene as the firefighters themselves.

The most common name for firehouse dogs was "Sparky," which then became the obvious name chosen by the NFPA for their fire prevention symbol.

Although horse-drawn fire engines in America faded from the scene during the 1910's and 20's, the tradition lives on. Dalmations, many of them still named "Sparky," are firehouse pets today, and some of them still ride on the apparatus to fires.

Smokey The Bear

The most successful public relations campaign ever in the United States to feature an animal, according to surveys, is Smokey the Bear, who is known not only to legions of children, but to virtually every American.

A cooperative nationwide wildfire prevention program was initiated in 1942, during World War II, with the involvement of state foresters, The Forest Service of the U.S. Department of Agriculture, and the Wartime Advertising Council. In 1945 they circulated a poster featuring Smokey the Bear, and the rest is history.

Since the campaign began, wildfires have dropped dramatically, including the total acres burned, and the lovable, yet authoritative Smokey Bear has been a huge success.

The slogan, "Only You Can Prevent Forest Fires," coined in 1947, has been so effective that it is still in use today, along with the added slogan, "Smokey's Friends Don't Play With Matches." Smokey's messages are taken very seriously by children, who remember Smokey's admonitions.

All 50 states participate in the Smokey Bear Junior Forest Ranger Program, as do Canada and Mexico, but south of the border Smokey has a different name. There he is "Simon El Oso," (Simon The Bear).

Yesterday's Leaky Fire Hose Was Made Of Leather

Just two centuries ago, fire hose as we know it today didn't exist in America because it was made of leather, and did it ever leak!

Stiff and unpliable, it had to be kept wound on a reel because to bend it sharply enough to flake it back and forth in a hose bed would have produced so many splits and cracks that little water ever would have made it as far as the nozzle. In fact, many firemen got soaked to the skin from the leaky hose, even though it was wound on reels.

The first real fire hose was invented in Amsterdam, Holland, in 1672 by cousins John and Nicholas Van der Heides, who fashioned it from leather sewn or stitched together with

The first real fire hose was invented in Amsterdam, Holland, in 1672 by cousins John and Nicholas Van der Heides, who fashioned it from leather sewn or stitched together with the stoutest thread they could find.

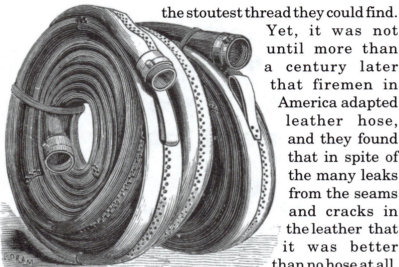

the stoutest thread they could find. Yet, it was not until more than a century later that firemen in America adapted leather hose, and they found that in spite of the many leaks from the seams and cracks in the leather that it was better than no hose at all.

Then, in 1808 came the big fire hose breakthrough. Members of the Philadelphia Hose Company introduced their new invention – leather fire hose held together with copper rivets. The rivets along the seams did what the best thread could not do – they held the hose together under pressure so that the leakage was drastically reduced.

Two members of the hose company – James Sellers and Abraham Pennock – with the approval of their fellow firefighters patented the idea and went into business manufacturing copper-riveted leather fire hose. Specifications for the newly invented hose were as follows:

Pure oak, city tanned, Baltimore or Philadelphia leather, known as "overweight" the average weight not less than twenty-two pounds to the side, none less than twenty pounds, double riveted with copper wire, size known as No. 8, twenty-two rivets to the running foot; splices made with thirteen rivets of size known as No. 7 wire, finished with three loops and rings and weight not less than sixty-four pounds to each fifty feet, exclusive of couplings, and warranted to stand a pressure of not less than two hundred pounds to the square inch.

The leather hose needed constant maintenance and had to be drained and dried after each use and then rubbed down with leather preservatives. Tallow, and neatsfoot oil obtained by boiling the hooves and shinbones of livestock, were favorites. Less favorite, though sometimes used, was smelly fish oil, which gave many firehouses of the day a characteristic aroma. These preservatives served to keep the hose pliable, preventing cracking and subsequent leakage.

It was not until fifty years later that leather hose was generally supplanted by cotton jacketed rubber hose. So, for a half century – from around 1820 to 1870 – the norm for fire hose was leather with copper rivets. Incredibly, many fire departments still had leather hose in service as late as 1900.

The Old Time Fireman's
Ultimate Disgrace

During the days of hand engine firefighting, no disgrace was attached to losing a building to the fiery element – which happened all too often. But volunteers dreaded one event more than any other – the humiliation of being "washed."

In those days – the early 1800's – each fire company was independent of any other. As a result, intense rivalry often arose between companies as to which one could get to the fire first, and which one could pump the most water. Since each company carried only 200 feet of leather, copper-riveted hose, it was often necessary to pump in relay. The first arriving engine set up operation at the nearest water source, which was usually a hydrant or cistern. They pumped to the next arriving engine, which in turn pumped to the third engine, then on to the fourth, fifth, and subsequent engines – as many as necessary to reach the fire.

One old time firefighter wrote of a relay of 30 engines – more than a mile long – to bring a single stream of water on a distant fire through a 7/8" nozzle. Relays with several engines were common, and 10 to 20 engines were not unusual.

The firemen often had a strong sentimental attachment for their particular engine, which always was

Firemen of New York City's Engine No. 14 show their dismay after having been washed by Engine No. 36.

given a name such as *Niagara*. When referring to their engine, the volunteers used the feminine gender, calling the engine "she" and "her." One New York City volunteer was once seen publicly hugging and kissing his engine after "she" had gotten the best of an engine from another company.

The volunteers pumped their hearts out, using as many as 150 strokes per minute, not necessarily deeply concerned with putting out the fire, but rather, concerned about not being "washed," meaning they couldn't pump hard enough to keep the water in their tank from overflowing onto the ground. If that happened, the firemen on the engine pumping into theirs would celebrate having "washed" them.

Being "washed" was a stigma that might never be forgotten nor lived down. Firemen who got "washed" were sick at heart, and many a fire laddie who would never shed a tear for any trouble of his own would break down and cry like a child if his engine were "washed" – the old time fireman's ultimate disgrace.

The Fire Engine Trust-Buster

Ever hear of American-LaFrance? Of course, because for an entire century – right up to the present – that name has been synonymous with the finest fire apparatus that money can buy.

The celebrated American-LaFrance entity came into being in 1902, when not only the American and LaFrance fire apparatus manufacturers merged, but nine others – a total of eleven manufacturers – came together to form what became widely known as "The Trust." Its purpose was not only to dominate but also to monopolize the fire apparatus industry in America by eliminating the competition.

The trust gobbled up nearly all the major manufacturers of both hand and steam pumpers, ladder trucks, chemical engines, and the only manufacturer of water towers. They were well on their way to achieving their goal of becoming a monopoly.

However, this trust business did not sit well with William S. Nott, an Irish immigrant who, starting from scratch, built a diversified manufacturing empire in Minneapolis. His factories turned out leather belting, rubber goods, including tires for new-fangled

W.S. Nott

automobiles, threshermens' supplies, pipe and boiler covering, and last but not least, hose carts and chemical fire engines. In addition, he acted as a distributor for fire apparatus manufactured by companies within the trust – until the trust revoked his distributorship.

William S. Nott was *not* a man to be trifled with. He resolved to show the trust a thing or two by running competition with a new line of fire apparatus he proposed to manufacture.

He hired H.E. Penney, the Minneapolis Fire Department's steam engineer and mechanic, to design and manufacture new steam fire engines to compete with the trust's "Continental" and "Metropolitan" models. This work was done in secret, inside a building Nott acquired for the purpose.

Next, also in secret, he put together a network of sales agents and turned them loose. By the time the first Nott steamer emerged from its hiding place, orders were in hand from Oklahoma, New Mexico, Pennsylvania, Montana, and even the Philippine Islands.

Nott undercut the trust's prices and offered steamers for as little as $3700, with a five year guarantee.

Soon he had more orders than he could fill, and he had to contract out some of the manufacturing process.

At first Nott steamers were sold mostly to smaller towns and cities, but soon orders began to arrive from Seattle, Los Angeles, Milwaukee, Minneapolis, Atlanta, New Orleans, and Chicago.

It appeared that Nott was indeed beginning to break the trust – at least as far as steamers were concerned. But beginning in 1910, something began to happen that not only ended the roll Nott was on, but quashed the steamer business for the trust as well.

It was simply new technology. Instead of ordering steamers from either Nott or the trust, fire chiefs across America were beginning to order new motorized pumpers – the kind that used gasoline-powered internal combustion engines, which were even more

A Nott steamer.

Photo from the collection of Ken Peterson

efficient than the tried and true steamers in use for half a century. The steamer business collapsed.

Did William S. Nott really break the fire engine trust? His biographer, Richard L. Heath, wrote "...[he] fought the 'fire engine trust' at the turn of the century – and won."

The Fire Alarm Tycoon
Who Hid In A Swamp

No inventor himself, John Nelson Gamewell, who is considered to be the father of fire alarm telegraphy, achieved dominance in his field by buying and exploiting the patents of others. As fast as new improvements in fire alarm telegraphy were invented, Gamewell was there to buy the patents and incorporate them into his ever-expanding line of fire alarm telegraph instruments.

By owning the patents and by using somewhat ruthless business tactics to squeeze out competitors, he built a fire alarm fortune. But in the early years, things did not always go well for the postmaster and telegraph agent from tiny Camden, South Carolina.

Without the capital to finance his first fire alarm venture, in 1855 he borrowed $30,000 to buy the patents of Dr. William F. Channing, a medical school graduate turned inventor. By 1861, Gamewell had installed

As fast as new improvements in fire alarm telegraphy were invented, Gamewell was there to buy the patents and incorporate them into his ever-expanding line of fire alarm telegraph instruments.

Over one million Gamewell fire alarm boxes were on America's street corners by the 1950's, nearly a century after their introduction.

fire alarm systems in Philadelphia, St. Louis, Baltimore, New Orleans, and Charleston, South Carolina. Then came the event that wiped him out financially and shattered his dreams – the Civil War.

Gamewell's showroom displayed various models of fire alarm boxes, alarm gongs, and other fire alarm telegraph instruments devised by inventors whose patents he acquired.

Camden, South Carolina, was in the direct path of General Sherman's march through Dixie to Georgia. Gamewell feared for his life, as he had joined the Confederate war effort as the operator of a facility making saltpeter, a gunpowder ingredient. He fled to a nearby swamp, where he remained hidden until Sherman had passed.

Unable to continue his fire alarm business during the war, Gamewell, by the time the South had surrendered, was so destitute that his brother-in-law loaned him five dollars. Meanwhile, the United States government had confiscated his patents and had sold them on the front steps of the Camden City Hall. It was a sad day for Gamewell, who was down, but not out.

He left South Carolina, moved to New Jersey, and persuaded investor John F. Kennard to buy back his confiscated patents. Kennard arrived in Washington prepared to pay as much as $20,000. Incredibly, he was able to purchase them all for the paltry sum of just $80.00!

Kennard and Gamewell formed a partnership and relaunched the business that went on to become a virtual fire alarm monopoly. By acquiring some of their competitors shops and forcing the rest out of business, they captured an estimated 95% share of the municipal fire alarm business in America.

"Fatal 48" - Louisville's Jinxed Fire Alarm Box

The calamities following alarms turned in from Buffalo, New York's, "Hoodoo" fire alarm Box 29 are well known to many fire buffs, but "Fatal 48," the box at the Louisville, Kentucky, corner of 6th & Main spelled death for even more firefighters than the box in Buffalo.

During the eleven year period from 1886 to 1897, 13 Louisville firefighters, as well as 10 civilians, perished at fires within a one block radius of Alarm Box 48. And that was only the beginning.

Around 11:00 p.m. on September 15, 1889, the watchman making his rounds along Louisville's Main Street saw a bright glow through the grating in front of Bamberger & Bloom Dry Goods Company. He ran to nearby Box 48, inserted his key, opened the door, pulled the hook once, and let go. Within the hour, five Louisville

Louisville firefighters were afraid of the box, and tried to break the jinx by switching the inner workings or movement of Box 48 with another box.

APPALLING!

Death and Destruction By Midnight Flames.

Awful Work of a Main Street Conflagration.

Brave Lives Lost In a Desperate Battle With Fire.

The Direful Calamity That Befel Louisville Last Night.

Five Gallant Firemen Killed and Valuable Business Houses Burned:

Bamberger, Bloom & Co.'s Store and Adjoining Buildings, Destroyed.

firemen who had responded to the box lay dead in the rubble of Louisville's deadliest fire ever.

Within a few more years, eight more firefighters responding to Box 48 perished, along with 10 civilians. By now, Louisville firefighters were afraid of the box, and they tried to break the jinx by switching the inner workings or movement of Box 48 with another box. Only the code wheel or "character wheel" with four teeth, a space, and then eight more teeth remained of the original mechanism in Box 48. But the substitution was for naught.

Soon the jinx struck again, with Box 48 calling still more Louisville firefighters to their deaths. And, with the turn of the century, the dreaded box was still sounding fatal alarms.

On May 6, 1908, Captain John Kreamer fell beneath the wheels of his responding hose wagon and was crushed to death. On September 23, 1911, three members of the Louisville Salvage Corps* met their deaths within a block of Box 48 when a wall collapsed

*See "The Men Who Wore These Helmets Weren't Firefighters" on page 77.

at the Baird Millinery fire on West Main. Then, on December 29, 1926, two more firefighters met their deaths at the corner of 5th & Main, just one block down from Box 48.

But "Fatal 48" will summon no more Louisville firemen to their deaths. All the fire alarm telegraph boxes in Louisville were finally removed in 1983, including the dreaded "Fatal 48."

The Musical Fire Alarm Code

Perhaps the most unusual audible code system for identifying the location of a fire was suggested (tongue-in-cheek, no doubt) back in 1835 by W.C. O'Brien of Osborn Hose No. 2, which protected a city whose identity is lost to posterity. Here is his unusual idea:

"I suggest having a clarion installed in the Court House Tower to ring automatically when an alarm is turned in, to play appropriate tunes so that firemen can tell at once the location of the fire."

Code and Location of Alarm Boxes

Thomson Street
  at the Cemetery .... *Nearer My God To Three*
Water Street .... *How Dry I Am*
Academy Street .... *School Days*
Catskill Point .... *On the Banks of the Wabash*
West Shore Depot .... *The Baggage Coach Ahead*
River Street .... *Where the River Shannon Flows*
William Street .... *Waltz Me Around Again Willie*
Greene Street .... *The Wearing of the Green*
King Street .... *All Hail the King*
Spring Street ...... *Apple Blossom Time in Normandy*
Bridge Street near the Jail .... *Prisoner's Song*

Division Street .... *West of the Great Divide*
Day Street ....*End of a Perfect Day*
Canal Street ....*Streets of Venice*
Grand Street .... *Down on the Rio Grande*
Old Ladies'
    Home .... *When You and I Were Young Maggie*
Church Street................ *Onward Christian Soldiers*
Garden Beauty .... *When Your Hair*
    Parlor   *Has Turned to Silver*
State Armory......................... *I Didn't Raise My Boy*
    Armory   *to Be a Soldier*

If a second alarm comes in
the chimes will play

"There'll be a Hot Time in the Old Town Tonight"

If the fire becomes an unmanageable
conflagration, the bells will chime

"Keep the Home Fires Burning".

The Piano Manufacturer Who Invented Automatic Fire Sprinklers

The inventor of the first widely used automatic fire sprinklers in the United States was piano maker Henry S. Parmelee, who was more concerned with his fire insurance premiums than with protecting his pianos.

As a result of disastrous conflagrations in Chicago in 1871 and in Boston in 1872, fire insurance companies increased their rates by a wide margin. As the premiums for Parmelee's New Haven, Connecticut, piano factory skyrocketed, he determined to find some way to lower them to their previous level.

If a fire in his factory could be detected automatically as soon as it started and, then, extinguished automatically, even before the fire department arrived, he reasoned that his fire insurance company would surely grant him a big discount.

He set to work by studying all the existing patents on automatic fire extinguishing systems and found that a British patent had been issued in 1723 using a network of fuses leading to charges of gunpowder. When a fire ignited a fuse, the resulting explosion would scatter water over the fire. Parmelee wisely rejected that idea.

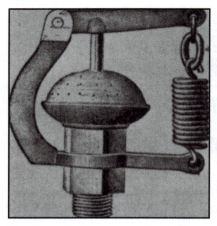

The Parmalee Sprinkler – In 1875, piano-maker Henry Parmalee of New Haven, Connecticut, developed the first sprinkler to receive widespread use. When intense heat melted the fusible link, the lever was raised, opening the water valve.

Then he found that a patent had been issued in 1806 for a network of perforated pipes with a water valve held shut by a weight on the end of a rope. When the rope burned through, the valve would open to fill the pipes – still not good enough.

Next, in 1812, a member of Parliment, no less, improved on the rope and weight scheme by substituting for the rope a cement that melted at 110°F to release the weight and open the valve. They were getting close, but Parmelee still wasn't satisfied.

He liked the fusible link idea, but instead of using it to open the main water valve, he installed one on each sprinkler head. That simple idea became the principle which ushered in a vast new fire protection industry. Within seven years after devising the fusible link sprinkler head in 1875, more than 200,000 were in use.

Today, installed throughout the world are multiplied millions of sprinkler heads which are based on the invention of the piano manufacturer who simply wanted to reduce his fire insurance premiums.

The Strange Fate Of The World's Largest Clipper Ship

Donald McKay had a dream. He envisioned a clipper ship twice as big as any other sailing ship ever built up to that time in 1853. It would be powered only by the wind blowing against enormous canvas sails on masts towering as high as a twenty-story building.

Investing all his personal fortune of $300,000, he laid the keel and hull at East Boston, Massachusetts. She was 325 feet long, 55 feet wide, and her rock-maple backbone was almost ten feet thick. The long iron bolts that held it together had to be pounded into place by a pile driver. An incredible 56 tons of copper bolts were used in its construction. So impressive was the monstrous sailing ship as it was being built that Longfellow was inspired to write a poem, which began:

> "Build me straight, O worthy Master,
> That shall laugh at all disaster..."

Christened the *Great Republic*, the hull slipped down the ways into Boston Harbor on October 4, 1853, as a band played "Hail Columbia," and cannons fired a salute. Donald McKay's dream was well on its way to fruition, but six more weeks of construction still lay ahead.

A model of the *Great Republic*, the largest clipper ship ever built, which was destroyed by fire on the eve of her maiden voyage.

Enormous iron-hooped, hard-pine masts stayed with ropes more than a foot in circumference supported 15,653 square yards of canvas sails. The paneled cabins were handsomely furnished. There was even a 36-foot-long velvet covered sofa installed.

A crew of 130 men and boys sailed her out of Boston Harbor, bound for New York, where she docked at the foot of Dover Street in Manhattan. She was loaded with 6000 tons of wheat and other cargo, worth over $200,000, which was destined for Liverpool, England.

The magnificent ship was set to sail on December 27, and although McKay was understandably excited, he fell sound asleep in his stateroom aboard the *Great Republic* just hours before the giant clipper was to cast off for England on her maiden voyage.

About 12:30 a.m. the cry of "Fire!" rang out over the east side of Manhattan. No, it was not the *Great Republic* that had taken fire – it was the Novelty Bakery, a full block away, that had been set afire from overheated ovens.

Soon flames shot through the bakery roof, and the fire spread to buildings on either side. On arrival, New York's Fire Chief Alfred Carson sounded a general alarm, bringing out nearly all the city's fifty hand-drawn fire engines.

But before most of them could get into action, the fire had spread to twelve more buildings, and the wind was pushing it toward the river. As it became apparent that the fire would indeed reach the docks, ferry boats and tugs began to pull sailing ships out of harm's way. But when they put their lines on the *Great Republic*, she wouldn't budge. The tide had gone out, and her heavy keel was resting on the bottom. Every tug in New York could not have moved her.

At the top of one of her masts – twenty stories above her deck – a spark lodged in the unfurled main topsail and set it on fire. Flames quickly spread to other sails. "There's just not anything we can do," Chief Carson told frantic Donald McKay. No hand fire engine ever built could push a stream of water that high. "Chief, can't you do so*mething*?" McKay pleaded. "I think you'll have to cut down the rigging," the Chief responded.

"I'll give a thousand dollars to any man who'll go aloft and cut off the fore and main masts above the masthead," McKay shouted to his crew. There were no takers – it would have been suicide.

Donald McKay, with his hands in his pockets and his head lowered, was a pathetic figure as he watched the flames consume his fortune and his dream on the eve of her maiden voyage. Fifty thousand days of labor were going up in smoke.

Although the firemen did their best to save the hull, it was not to be. Thousands of tons of wheat were now burning in the hold, immune to the many hose lines pouring thousands of gallons of water each minute. Finally, deck hands opened holes below the waterline and sunk what was left of the *Great Republic*.

Even the hull could not be salvaged. The water-soaked wheat had swollen and burst the hull beyond repair. The ship and her cargo were a total loss. Close friends said McKay aged twenty years that night. Insurance covered less than half the loss. Donald McKay was a broken man, with both his fortune and his dream forever gone.

He eventually retired to a farm in Massachusetts, and on the broad mantel in his den rested a model of the *Great Republic*. It was said that often, at the close of day, he would sit quietly and look up at her, perhaps fulfilling his dream the only way now possible – in his imagination.

George Washington's Last Alarm

George Washington, who was our first president of the United States, and who is often considered to be our nation's greatest hero, became an active volunteer fireman in 1750, at age 18, with the fire company in Alexandria, Virginia, even though he lived several miles away.

He was born in 1732 on a small, struggling tobacco farm in Virginia. His father died when he was 11, and he had to go to work to help his family make ends meet. By age 15, he was working as a professional surveyor beyond the Blue Ridge Mountains far away from home.

But at age 18, we find him residing with his older brother Lawrence at Mount Vernon. He made the trip into town on horseback as often as 10 times a week, and as an Alexandria fireman, he responded to fires on horse- back from Mt. Vernon.

George Washington riding to a fire.

George Washington as a fireman.

The Alexandria firefighters had no engine, and used only leather buckets to fight fires. They also carried bags made from "oznaburg" or "widder linnen" to carry out valuables from burning buildings, which often burned to the ground in spite of the valiant efforts of the bucket brigade.

By 1774 Washington was a member of the Continental Congress at Philadelphia, and he had no time for fighting fires back in Virginia. But the Friendship Fire Company of Alexandria honored their now famous former comrade by making him an honorary member of the fire department, and they sent him a copy of the minutes showing that the honor had been bestowed on him by the unanimous vote of the membership.

This pleased Washington so much that he promptly inspected all the fire engines in Philadelphia to determine which type was the best. He selected an engine built by a man named Gibbs, paid eighty pounds and ten shillings for it, and just before he set out for Boston to become Commander-In-Chief of the Continental Army, he sent the engine as a gift to his friends in Alexandria.

George Washington never lost his interest in the fire service and would often visit fire departments during his travels around the country. After his second term as President of the United States, he retired to Mount Vernon and, once again, frequently visited nearby Alexandria.

In 1799, the final year of his life, when he was 67 years of age, he was one day riding down King Street in Alexandria when a fire broke out near the market. Arriving at the fire, he saw that the Friendship engine was poorly manned, although a number of

Alexandria firemen used only leather buckets to fight fires.

well-dressed spectators were gawking at the struggling volunteers.

Angered, Washington rode through the crowd shouting, "It is your business to lead in these matters!" and throwing the bridle of his horse to his servant, he leaped to the ground, grabbed the engine's pump handles himself, and began to vigorously work them up and down. Soon the onlookers also grabbed onto the handles, and the pressure in the hoses increased dramatically.

The market was saved, but it was George Washington's last alarm. Before the year had drawn to a close, the Father of our Country, who had fought his first fire at age 18, and his last fire nearly a half century later at age 67, had passed on to a hero's final reward.

The Ghost Of The London Fire Brigade Museum

You may not believe in ghosts, but as for John Rodwell, Curator of England's London Fire Brigade Museum, he believes. Here is his story as it appears in the London Fire Brigade Museum Guide, and as it was related to the author by Mr. Rodwell during a visit to the museum.

The London Fire Brigade Museum is located in Winchester House – an imposing structure housing one of the most comprehensive collections of firefighting equipment and memorabilia in England. It was built in 1820 on Southwark Bridge Road on the site of the graveyard of St. Saviour's Workhouse. According to the guidebook, "It is said that the spirits of those buried beneath return to haunt the rooms if disturbed."

> *"It is said that the spirits of those buried beneath return to haunt the rooms if disturbed."*

They were greatly disturbed in the 1980's when workmen renewing the parking area outside the museum dug down for the foundations and came

The author holds an eighteenth century brass "squirt" in the ghost's favorite room at the London Fire Brigade Museum.

across the bones and skulls of workhouse inmates. In those days they never had the money to give those unfortunates a formal burial, so they just dug a hole in the ground and threw the bodies in.

When they discovered the bones and skulls, and while the grave was open, the museum began to experience all sorts of strange happenings. Helmets were taken off shelves by unknown forces and thrown across the room. Paintings were lifted off the walls and thrown about, breaking their frames. Although electronic burglar alarms were installed, the mischief continued without the alarms ever being tripped, which, according to the alarm company, was impossible.

There is a ship's bell on the first floor landing of the museum, and on one particular occasion, when the curator was on the first floor and his assistant was on

London's prestigious Fire Brigade Museum is said to be haunted.

the top floor, the bell began to ring. The curator shouted upstairs to his assistant, "Visitors are about to arrive. Will you please stop ringing the bell!" "I'm on the top floor. I'm nowhere near the bell!" came the reply. There was no one else in the museum.

The bell didn't stop ringing until the curator got halfway up the stairs. That in itself was mysterious, but weirder still was the fact that because school children often rang the bell as they walked past, the curator had the clapper removed. It was now impossible for the bell to ring.

After the authorities had examined and recovered the grave, they had the museum exorcised by priests, after which the strange happenings ceased. But even so, sometimes visitors to the museum, today, say they can smell an odd smell and that they feel a chill in certain rooms as they walk through the museum.

The Fire Alarm Booth
That Trapped You Inside

It's a pleasant Sunday evening, and as a law-abiding citizen, you are strolling down the street minding your own business. You pass a display window that you notice is filling with smoke. Doing your civic duty, you run to the fire alarm booth at the corner, enter, shut the door, and push the alarm button.

Then you hear an ominous click. When you try to open the door to leave, you can't, because you're locked inside!

Responding to the alarm will be not only the fire department but also the police patrol wagon. A policeman will set you free, then commend you for turning in the alarm for a real fire or arrest you for sending a false alarm.

Did it really happen? Probably not, as there is no record of a single culprit-trapping

In those days, the boxes were kept locked to avoid false alarms, and a person trying to send an alarm first had to locate the key, usually held by the policeman on the beat.

booth ever having been installed, but it wasn't because the Gaynor Electric Company of Louisville,
Kentucky, didn't try.

Gaynor, before the turn of the century, was a
competitor to the celebrated Gamewell Company,
both of whom manufactured the fire alarm telegraph
boxes mounted on street corners. In those days, the
boxes were kept locked to avoid false alarms, and a
person trying to send an alarm first had to locate the
key, usually held by the policeman on the beat.

Excerpts from Gaynor's advertising literature for
their lock-you-inside alarm booth included the following:

It is a well-known fact that considerable time
is always lost after the discovery of a fire until the
alarm can be turned in from the box. How to
reduce this interval, which always amounts to
several minutes, has heretofore remained an
unsolved problem. This difficulty has been entirely overcome by the invention of our combination Fire and Police Signal Station, which con-
sists of an octagonal iron booth, the door of
which is always unlocked. It is only kept closed
by the aid of a spring and knob latch, the latter
being on the outside of the door.

Any person upon the discovery of a fire can
enter this Station, and after having closed the
door, will, by pushing a button, give the alarm of
fire. It can be readily seen that by means of this
device the fire department can be brought to the
Station ready for action long before the alarm
could have been given, if the discoverer of the fire

A policeman releases the prisoner in the fire alarm booth as the fire department arrives. It's not a false alarm, but a real fire. Note the smoke and flames at the upper right hand corner of the picture.

had first to procure a key from some police officer or other key holder.

To prevent the giving of false alarms, the door of this Station is provided with an electric lock so that the person giving the alarm is placed under detention. This arrangement being once understood, the moral effect alone will prevent the giving of false alarms by persons maliciously inclined.

Gaynor's invention was never accepted, not even in his home town of Louisville, Kentucky, and soon the company found itself out of business.

Sources

The Steamboat Fire Whose Death Toll Exceeded The *Titanic*
Elliott, J.W. *Transport to Disaster*
Berry, Chester *Loss of the Sultana and Reminiscences of Survivors*
Potter, Jerry O. *The Sultana Tragedy: America's Greatest Maritime Disaster*
Salecker, Gene *The Crippling of the Sultana*
Deason, Mike *Memoirs of Jesse Marshall Huffaker (A Sultana Survivor)*

The Fire Inventor Whose Tanks Helped Win World War II
Christie, J. Edward *Steel Steeds Christie*

America's Foremost Female Firefighter
Holdredge, Helen *Firebelle Lillie*

The Day An Airplane Hit The Empire State Building
Weingarten, Arthur *The Sky Is Falling*

The Confederate Plot To Burn New York
Brandt, Nat *The Man Who Tried To Burn New York*

The Civil War Zouaves
Costello, Augustine E. *Our Firemen*
Morris, John V. *Fires and Firefighters*
Smith, Dennis *History of Firefighting in America*

Motorcycles That Fought Fires
Dodson, Jim *Indian Fire Patrol*

The Fire-Fighting Trolley Car
Goodman, M.W., M.D. *Inventing The American Fire Engine*

The Fire That Made Beggers Rich
Ditzel, Paul *The Day Bombay Blew Up*

The Circus Fire Mystery
Ditzel, Paul *Firefighting During World War II*

Colonial America's Foremost Fire Engine Builder – An Apprentice Of Paul Revere
Tufts, Edward R. *Hunneman's Amazing Fire Engines*

The Fire Department Without A Pumper
Weaver, Betsy; and Frederick, Gary E. *Hands, Horses and Engines*

A Thing Of Terror
The Fireman's Herald
Amoskeag Steam Fire Engines
The Famous Amoskeag Steamers

The Men Who Wore These Helmets Weren't Firefighters
Douglass, Emmons E. *While The Flames Raged*

The Day Pennsylvania's State Capitol Burned
Houseal, Robert M. *The Pennsylvania State House Fire*

The Suspicious Cruise Ship Fire That Drew One Million Spectators
Burton, Hal *The Morro Castle*
Gallagher, Thomas M. *Fire at Sea; The Story of the Morro Castle*
Taggart, William S. *The Morro Castle Disaster*

The *Hindenberg* Disaster And The Little Engine That Couldn't
Mooney, Michael *The Hindenberg*
Ronan, Margaret *The Hindenberg Is Burning*

Yesterday's Triggerless Dry Powder Extinguishers
Pajak, Jerry *What's Old Is New Again*
Hirsch, Dave *"Past Times"; Fire and Other Extinguished Advertising*

Breaking The Fire Lantern Color Code
Pajak, Jerry – *seminar at the Fire Mark Circle Convention*
Rosenblum, Harry *Lantern Globes*

Fire Marks

Maher, Thomas M. *The Many Adventures of
Fire Mark Collectors*
Atwood, Becky *Missouri Historical Society*
Shea, Bob *Fire Marks – Part of Insurance History*

Ben Franklin's "Ounce Of Prevention" Led To Volunteer Firefighting In America

The Philadelphia Contributorship For The Insurance
Of Houses From Loss By Fire *Franklin & Fires*

Did Saboteurs Really Torch The Luxury Liner Normandie?

Ditzel, Paul *Firefighting During World War II*

Japanese Incendiary Bombs That Fell On The United States

Ryczkowski, John J. *World War II Japanese
Bombing Balloons [in American Fire Journal]*

The Underwater Fire Station

Rose, Robert L. *These Offices Operate Underwater (Really)
[in the Wall Street Journal]*

The Grand Masters Of Vintage Fire Engine Restoration

Dickinson, Regan D. *My Favorite Part Is Decorating
[in Sign Business]*
Weber, Tom *A Fire Engine Affair
[in the Bangor Daily News]*
Kiefer, Kathy *Dreams of Gold [in Signcraft]*

The Kitty Litter Fire That Extinguished Itself

The Portland, Maine, Press Herald

Fire Service Traditions

Foley, William R. *The Book of Florian*
Forest Service, U.S. Department of Agriculture
Fire Almanac
Cork County Centenary Review

Yesterday's Leaky Fire Hose Was Made Of Leather

Costello, Augustine E. *Our Firemen*

The Old Time Fireman's Ultimate Disgrace
Costello, Augustine E. *Our Firemen*

The Fire Engine Trust-Buster
Heath, Richard L. *Fighting the Fire Engine Trust: The Nott Fire Engine Company of Minneapolis [in Minnesota History]*

The Fire Alarm Tycoon Who Hid In a Swamp
Ditzel, Paul *Fire Alarm!*

"Fatal 48" - Louisville's Jinxed Fire Alarm Box
Abner, Rhonda *Louisville Fire Department History*

The Musical Fire Alarm Code
Thanks to Fire Historian David Lewis

The Piano Manufacturer Who Invented Automatic Fire Sprinklers
Greer, William *A History of Fire Alarm Security*

The Strange Fate Of The World's Largest Clipper Ship
Morris, John V. *Fires and Firefighters*

George Washington's Last Alarm
Costello, Augustine E. *Our Firemen*

The Ghost Of The London Fire Brigade Museum
London Fire Brigade Museum Guide

The Fire Alarm Box That Trapped You Inside
Thanks to Robert W. Fitz for supplying the Gaynor catalog

Firefighting History VIDEOS

Produced and Narrated by W. Fred Conway

Video —
Those Magnificent Old Steam Fire Engines

Produced for cable television, this documentary program includes spectacular footage of restored steamers in action as well as vintage footage shot in 1903 from the archives of the Library of Congress. VHS, 40 minutes.

- You'll watch two terrifying steamer accidents caught live on-camera:
  - A video cameraman narrowly escapes serious injury when the steamer he was filming explodes and blasts live steam at the spot where he had been kneeling just moments before. His camera kept rolling.
  - You'll watch a second thriller when the big air chamber of another steamer blasts off like a rocket and goes flying through the air.
- You'll watch the "old-timer" recount his vivid recollections of steamers when he was a lad.
- You'll see a water tower raised by hydraulic pressure supplied by a steamer, and watch it blasting water pumped by two steamers.
- You'll watch a fire alarm coming in to a 100-year-old alarm office with the dispatcher transmitting it to fire stations with steamers and horses.
- Narrated by W. Fred Conway and Stephen Heaver, Jr., curator of The Fire Museum of Maryland.

$19⁹⁵

Video —
"The Old Machine"

The Incredible Rebirth of an American-LaFrance Metropolitan Steam Fire Engine.

"The Roosevelt" was hidden away in an obscure museum annex when Fire Lieutenant Brent Palmer discovered it and quit his job in order to devote his life to restoring it to its original splendor — and then parading and pumping it dozens of times each year.

Follow Brent's incredible restoration of more than 22,000 individual parts as he recreates the magnificent splendor of one of the most stunning fire engines ever manufactured — the American-LaFrance Metropolitan steamer.

$19⁹⁵
